The Origin of Higher Taxa

T0138207

The Origin of Higher Taxa

Palaeobiological, developmental, and ecological perspectives

T. S. Kemp

OXFORD
UNIVERSITY PRESS

and

The University of Chicago Press
Chicago and London

OXFORD
UNIVERSITY PRESS

Great Clarendon Street, Oxford, OX2 6DP,
United Kingdom

Oxford University Press is a department of the University of Oxford.
It furthers the University's objective of excellence in research, scholarship,
and education by publishing worldwide. Oxford is a registered trade mark of
Oxford University Press in the UK and in certain other countries

© T. S. Kemp 2016

The moral rights of the author have been asserted

First Edition published in 2016
Impression: 1

All rights reserved. No part of this publication may be reproduced, stored in
a retrieval system, or transmitted, in any form or by any means, without the
prior permission in writing of Oxford University Press, or as expressly permitted
by law, by licence or under terms agreed with the appropriate reprographics
rights organization. Enquiries concerning reproduction outside the scope of the
above should be sent to the Rights Department, Oxford University Press, at the
address above

You must not circulate this work in any other form
and you must impose this same condition on any acquirer

Published in the United States of America by the University of Chicago Press
1427 East 60th Street, Chicago, IL 60637, United States of America

British Library Cataloguing in Publication Data
Data available

Library of Congress Control Number: 2015940142

ISBN 978–0–19–969188–3 (hbk, Oxford University Press Edition)
ISBN 978–0–19–969189–0 (pbk, Oxford University Press Edition)

ISBN 978-0-226-33581-0 (hbk, The University of Chicago Press Edition)
ISBN 978-0-226-33595-7 (pbk, The University of Chicago Press Edition)

ISBN-13: 978-0-226-33600-8 (e-book, The University of Chicago Press Edition)
DOI: 10.7208/chicago/ 9780226336008.001.0001

Library of Congress Cataloging-in-Publication Data

Kemp, T. S. (Thomas Stainforth), author.
 The origin of higher taxa : palaeobiological, developmental, and ecological perspectives /
Tom Kemp.
    p.   cm.
 Includes bibliographical references and index.
  ISBN 978-0-226-33581-0 (cloth : alk. paper) — ISBN 978-0-226-33595-7 (pbk. : alk. paper) —
ISBN 978-0-226-33600-8 (ebook) 1. Biology—Classification. 2. Evolution (Biology) I. Title.
 QH83.K46 2015
 578.01′2—dc23

                                                                  2015008533

Printed and bound by
CPI Group (UK) Ltd, Croydon, CR0 4YY

Links to third party websites are provided by Oxford in good faith and
for information only. Oxford disclaims any responsibility for the materials
contained in any third party website referenced in this work.

For Małgosia with love and thanks as ever

Preface and Acknowledgements

I have spent a fair proportion of my academic lifetime thinking, writing, and teaching students about how fossils can contribute to evolutionary theory. In essence, looking into how to handle the apparent circularity whereby on the one hand a pre-existing theory of evolution is needed in order to give meaning to the scraps of preserved ancient anatomy called fossils, while on the other hand the vast timescale of the fossil record reveals aspects of long-term evolution neither discernible nor predictable from the study of living organisms: between evolutionary process as cause and phylogenetic pattern as effect. What I am sure about is that resolution lies in the properly applied combination of evidence about the nature of living organisms drawn from functional biology, genetics, and ecology with that from palaeobiological observations about actual evolutionary change. An earlier work of mine, *Fossils and evolution* (Oxford University Press, 1999), was explicitly predicated upon such a viewpoint, and the present work is an expansion of one of its chapters. It concerns the forces that cause and direct occasional evolutionary lineages to evolve to such an extent that they culminate in what is recognized as a new higher taxon. The paradigm of this phenomenon for me has always been the origin of mammals, but I believe that the generalized conclusions reached in that case can be extrapolated with illuminating effect to those other cases where the fossil evidence is less revealing. I am sure that the questions I pose about evolution at this scale are scientifically valid, important, and of course extremely interesting ones, even if scandalously neglected in recent times by evolutionary biology. Undoubtedly I have not successfully identified all the right answers. However, I am equally convinced that the opinion of some evolutionary biologists that all evolution can be adequately explained by simple intrapopulational selection alone is not defensible, and that this book is an essential counterpoint to that point of view.

I am indebted to colleagues who have read and commented on various of the draft chapters: Per Ahlberg, Mike Benton, Andrew Berry, Jessica Bolker, Adrian Friday, Alan Love, and Derek Siveter. All made very helpful suggestions that I have incorporated, and I apologise to them for the occasions I did not act upon their comments when, always in the most courteous of fashions, they did not entirely agree with some of my interpretations. I absolve them from any responsibility for what remains.

I thank the numerous publishers who have given permission for the reproduction of copyrighted illustrations: individual acknowledgement of sources are indicated in the captions. Attempts have been made to contact all copyright holders where relevant: if anyone has any queries regarding the reuse of their material, please contact the publisher.

I should like to thank the OUP editorial and production team of Ian Sherman, Lucy Nash, Viki Mortimer, and Janet Walker for their patience and encouragement. The synergy between author and publisher is rather like that between fossils and evolution: neither is paramount and the outcome is more than the sum. I would also like to thank Christie Henry and Amy Krynak from Chicago University Press.

Finally I am deeply grateful to St John's College, Oxford for electing me to an Emeritus Research Fellowship to allow me to continue writing since my retirement in comfortable and congenial surroundings.

Tom Kemp
St John's College, Oxford
Autumn 2015

Contents

CHAPTER 1

Introduction

1.1 The question

The question I address in this book is how new higher taxa evolve, taxa that can be described as the culminations of long treks through a multidimensional morphospace (Fig. 2.2) during which many characters change and novel characters appear. I come to this as a vertebrate biologist with a lifetime of interest in palaeobiology and what fossils can tell us about the processes of evolution involved. My particular research has largely concerned the origin of the mammals, which is associated with the most detailed and therefore most informative fossil record of the pattern and sequence of acquisition of new characters of a major transition. To a varying degree, a number of other higher taxa of vertebrates, notably tetrapods, birds, turtles, and whales also have fossil records of their stem members. The fossil record is a great deal less complete and its interpretation more controversial with regard to the origin of the invertebrate phyla and classes. My aim is to explore whether there are processes and circumstances that apply generally to the origin of new higher taxa. However, the fossil record is only a part of the evidence bearing on this quest. Indeed we can only hope to understand the morphology, palaeoecology, and inferred evolution of fossils in the light of the general nature of organisms as complex systems embedded in real environments, and which possess the genetic and developmental mechanisms capable of change that we know of from modern organisms.

Perhaps I can put the question thus: are major evolutionary transitions adequately accounted for by normal Darwinian natural selection proceeding for a sufficient length of time, or are unusual genetic processes and/or special environmental circumstances required? The default explanation, expressed implicitly if not explicitly by the vast majority of evolutionary biologists, is that natural selection acting on an interbreeding species population is sufficient explanation, and that the particular genetic changes and environmental conditions are of no special significance beyond the contingencies of each individual case. However, the very act of making this reductionist prior assumption universal leads to the automatic exclusion from consideration of other possible evolutionary processes that might exist but that are too slow and too long-acting to be observable in living populations. It is therefore necessary that we consider all the evidence available bearing on the origin of higher taxa to see if any such extensions of evolutionary theory are justified. In fact, study of evolution at this level, represented by the origin of, for example, the invertebrate phyla and classes, and vertebrate taxa such as teleosts, turtles, birds, and mammals, and perhaps the families of flowering plants, has been seriously neglected. In part this is perhaps because the field has often been discredited in the past by non-scientific, vitalistic concepts of instant transformation and internal evolutionary drives. Even at best it has often been belittled as the stuff of nothing more than imaginative speculation about individual cases beyond practical testability, of interest only in the museum exhibition or popular book. I believe this neglect is a shame because asking, for example, what kind of mechanisms and circumstances caused an ancestral worm-like creature eventually to give rise to a lobster or squid, or a cold-blooded, scaly, sprawling-limbed, tiny brained amniote to end up as a mouse or a bird, are certainly rational questions: we have no reason to doubt that these events happened and therefore that they are in principle amenable to scientific inquiry. Furthermore, they are extraordinarily

The Origin of Higher Taxa, First Edition. T. S. Kemp.
© T. S. Kemp 2016. Published 2016 by Oxford University Press and The University of Chicago Press.

important and interesting questions if explanations are to be sought for the diversity of life on earth.

An observation so commonplace as to be scarcely ever remarked upon is that the organisms alive today tends to fall into discrete clusters of those whose morphology is relatively similar, but which differ more markedly from the morphology of the members of other such clusters. This is true at every level in the taxonomic hierarchy: a typical genus is such a cluster of relatively similar species, a family is a cluster of genera, and so on right up to the level of phyla as clusters of classes. Moreover, the present day biota is a single time-slice. Although it is the one about which by far the most is known, it is not actually different in principle from earlier time-slices. Looking, for example, at the Upper Permian period about 260 million years ago, an equally clear-cut hierarchy of discrete taxa presents itself in the fossil record, including many, like trilobites and pareiasaur reptiles, that have long since disappeared. Why should this pattern of taxa be generally so? Why do taxa not form a continuum, or at least a more even distribution in morphospace? It could be argued that the hierarchical pattern of kinds of morphology reflects a comparably discontinuous environment to which taxa have to adapt, but the environment is in fact a good deal more continuous than the kinds of organisms that inhabit it. Most of its variables, such as mean temperature and humidity, salinity, altitude, day length, vegetation cover, food particle size, and so on, form gradients. Alternatively the hierarchical pattern of morphology may reflect a tendency for the evolutionary processes underlying speciation and extinction to cause lineages to diverge from one another. But if so, what is it about the processes that control taxonomic turnover to such a degree that the lineages diverge sufficiently to end up being regarded as different higher taxa?

1.2 The context

The vast bulk of research on evolutionary theory concerns evolutionary change at the low taxonomic level and relatively very short timescale referred to as microevolution: evolution within a continuous interbreeding population. Much indeed has been written about the most fundamental unit step in evolution, which is the origin and fixation of genetic mutations by selection, or more passively by genetic drift and hitch-hiking, in the population. At the phenotypic level, a large part of evolutionary biology has concerned intraspecific variation and associated selection forces acting on single morphological characters such as the size of birds' beaks, colour patterns associated with predation and industrial melanism, and many other examples. At this microevolutionary level, the variables are sufficiently few and simple for the evolutionary process to be mathematically tractable and readily modelled. The timescale involved is sufficiently short for hypotheses to be testable by empirical observations of real cases in the field or laboratory. In this way, vast knowledge has accumulated about the genetic and ecological principles of natural selection, and in more recent years about the molecular basis of developmental gene function and how it effects phenotypic change.

Beyond this level of analysis, things grow more difficult and indirect. It is at this point that evolutionary biology transforms from an essentially experimental science to an historical one. Any process that takes more than a few decades to complete is not amenable to complete observation or experimental manipulation, and consequently other methodologies and sources of evidence are needed. These include predictions of evolutionary consequences inferred from the genetic and ecological natures of living organisms under assumed environmental conditions, and often embrace mathematical and computer-modelling approaches. A more empirical approach consists of inferences about the whole process from observations of presumed intermediate stages frozen in the time-slice of today. Speciation is the most common case studied by such means. It is an evolutionary process whose time course is typically (though with huge variation) measured in thousands of years, which is vastly longer than is available for direct observation. Consequently, most theories of speciation are derived from theoretical modelling of what would be expected to happen to a gene pool subjected to various patterns of biogeographic distribution and regimes of selection forces. Empirical work on speciation is mainly a matter of indirect inference from snapshots of examples of population divergence actively in progress or recently completed, such as clines, superspecies, sibling species, and hybridization. It is true that some conclusions have been drawn from experimental work involving artificially imposed selection forces on laboratory populations leading to reproductive isolation, but how representative of the actual complexities of natural evolution such simple and controlled studies are is, to say the least, debatable.

At an even greater timescale still, the fossil record starts to provide direct evidence of the outcome of evolutionary processes. However, with rare exceptions, its temporal resolution is at best of the order of 10^4–10^5 years. This means, for example, that if a speciation event occurred which took as much as 10,000 years from inception to completion, on average not a single intermediate stage between the ancestral and the derived species populations would be preserved, a limitation exacerbated if the speciation event happened to involve a small, geographically restricted population. Certainly, for all but a tiny handful of the vast majority of speciation events over geological time no intermediate stages at all are known. Thus the fundamentally important evolutionary process of speciation is too slow for direct observation but too fast for its course to be recorded in the fossil record. It falls into what has been termed the 'epistemological gap' (Kemp 1999). No wonder that discovering the actual processes and sequence of events in real cases of speciation has proved so difficult.

Going beyond the speciation level is to enter the realm referred to as macroevolution, or transspecific evolution as it has been referred to by some. Here the main concern is to explain patterns of taxonomic diversification and turnover, as revealed indirectly by the systematics of modern organisms, and directly by the fossil record within the inevitable constraint of its high incompleteness. Estimates are that only perhaps 1 per cent of all potentially fossilizable species are actually known: for example, *The Paleobiology Database* currently lists about 135,000 species which, assuming on highly speculative grounds that 13.5 million macroscopic species with mineralized skeletons have existed, gives this proportion (see <https://paleobiodb. org>). The macroevolutionary level of study is entirely remote from direct empirical observations

of evolutionary mechanisms. Only the outcome is available as evidence, in the form of the description of taxa existing today, or of the appearances and disappearances of taxa in the fossil record. As far as the modern biota is concerned, the major foci of macroevolutionary study include such aspects as the relative species diversity of particular genera and families, and inferences about the forces of natural selection in the past that may have driven this as habitats were increasingly subdivided into more and more niches. The taxonomic level studied varies from the species such as those of *Drosophila* on the Hawaiian islands, through genera, for example the celebrated geospizid finches on the Galapagos Islands, to families and orders such as many investigations of mammals.

Considering the fossil record, due to its incompleteness it is more practical and less prone to sampling error to analyse patterns of turnover of taxa above the species level, usually genera or even families. Most of the work in macroevolution by palaeobiologists concerns such issues as measuring changing rates of diversification and extinction of taxa over geological time, including the spectacular mass extinctions when as many as 90 per cent of species were lost, and how these relate to palaeo-environmental and palaeogeographical factors.

The occasional origin of new higher taxa, the subject of this work, can be viewed as a special aspect of the evolutionary pattern, one that is embedded within the overall taxonomic turnover revealed by fossils. Despite the relative paucity of research, this is a field of the greatest fundamental importance to biology. How can evolving lineages trek long distances through morphospace without the succession of phenotypes that constitute them ever losing the functional integration they must possess as viable organisms? How can and why do lineages become so distinct from their contemporaries even if they commenced from the same common ancestor? What combination of environmental and internal conditions drives lineages in the directions they happen to take?

1.3 The available evidence

The following chapters review the different categories of evidence that are pertinent to a search for explanations of the origin of new higher taxa. Each category is relevant, none is treated as paramount, and the attempt to reach a synthetic conclusion of broad application depends on combining all.

Chapter 2 is a consideration of what kind of entity higher taxa are, such that they can be recognized and defined, however arbitrarily. They may be viewed respectively as formal systematic taxa at a certain level in a Linnean hierarchy; as groups of species occupying a particular volume of hyperdimensional morphospace, and as groups of organisms occupying a particular peak or group of closely spaced peaks on an adaptive landscape. Each of these perspectives offers initial insights into how they may originate and evolve.

The most fundamental kind of evidence consists of inferences from the very nature of organisms. As discussed in Chapter 3, when an organism is viewed as an integrated system of functionally interrelated parts rather than as a taxonomist's list of characters, constraints are imposed on how it could have evolved from its remote ancestor. Exploring organisms from this perspective leads to insights into how such a system over time could accumulate very large evolutionary changes in many traits, yet remain throughout a fully integrated, adequately adapted phenotype. This approach generates several concepts which may be applicable generally, but it does not provide empirical evidence about what actually happened in particular cases of identified higher taxa. For this it is necessary to turn to other sources.

The most direct empirical evidence is the fossil record, the nature of which is discussed in general terms in Chapter 4. It is intermediate-grade fossils, fossils that combine some of the characters of the ancestor with some of the derived characters of the higher taxon evolved from it, that are of interest, although almost invariably these fossil taxa also possess unique characters relating to their own special adaptations and are therefore ineligible as candidates for direct ancestry of the higher taxon. However, as explained in the chapter, this is unimportant as far as reconstructing hypothetical intermediate morphologies is concerned because these specialized, or autapomorphic characters can be ignored.

As discussed in Chapter 5, embryological evidence has always played an important role in evolutionary thinking, both in the context of establishing phylogenetic relationships and as a source of hypotheses about evolutionary mechanisms of change. The naïve recapitulation theory of Haeckel and its subsequent modification to the more general idea of an increasing similarity between earlier stages of different developing embryos has figured extensively in early phylogenetic studies. As for mechanisms, the phenomenon of heterochrony, whereby the relative timing and rate of development of different parts of an organism can be altered, points to an important mechanism for integrated morphological change that can be manifest at a relatively high taxonomic level. So does the related idea of heterotopy, described as shifts in the relative position at which structures develop. Allometry is where different parts of the organism bear different size relationships to overall body size, so if the latter changes during evolution, then proportions amongst body parts also change in a coordinated fashion.

The molecular revolution in biology has touched virtually every field, and this one is no exception, as more and more is learned about the molecular basis of how genes control embryological development. By discovering correlations between differences in gene sequences and the pattern of expression of genes on the one hand, and phenotypic differences on the other, it becomes possible to begin opening up the 'black box' that lies between specific mutations and particular phenotypic changes in evolved lineages. This subject of molecular 'evo–devo' has at present a long way to go before a useful general theory of major evolutionary transition at the molecular level emerges, but there are nevertheless numerous smaller-scale examples pointing towards the general principles, such as cis-regulation and hierarchically organized genetic developmental networks. Chapter 5 includes a necessarily sketchy overview of this rapidly expanding area of evolutionary biology.

Every species that has lived is assumed to have been adequately adapted to its niche in the environment, an assumption that includes all the intermediate species within an evolving lineage, however much it evoloves. Chapter 6 considers the nature of the ecological settings associated with major transitions, in an attempt to account for the direction and pace of change leading to a new higher taxon. This is perhaps the most elusive category of actual evidence. Modern populations and communities are subject to ecological processes that are generally understood, but only over the very short term. Information about the geological record provides a certain amount of palaeoecological information associated with important phases of taxonomic turnover such as mass extinctions and extensive radiations, but this is very incomplete and coarse-grained at best. The ecological circumstances surrounding the origin of new higher taxa is extremely limited by the incompleteness of the palaeoecological evidence, and in any case it does not usually relate to the actual transitional taxa since the latter are almost never known as fossils anyway. Therefore hypotheses concerning the ecological driving forces acting on the sequence of hypothetical ancestors and descendants rely on inferences from a range of particularly indirect and incomplete empirical evidence. Nevertheless, an attempt is made to seek general principles, given the pivotal role the environment plays.

The actual fossil record of the stem groups of all those higher taxa where relevant material exists is reviewed in Chapters 7 and 8. Disappointingly, few such higher taxa possess much in the way of stem-group fossil members. In the case of the invertebrates, there are several early fossils that have been proposed as intermediate in grade between a common ancestral metazoan and one or another derived phylum. However, there is often deep disagreement about the anatomical interpretations and even, in some of the most critical cases, about which phylum a particular fossil is actually a stem member of. Amongst the vertebrates, with their altogether greater number of fossilizable parts and their more recent history, there are some much more informative examples, notably concerning the origin of tetrapods, mammals, and birds, plus at least some information on several other distinctive higher taxa. These offer direct information about the sequence of acquisition of the derived characters of the higher taxon in question, from which inferences can be drawn about the processes of evolution involved.

1.4 Synthesis

How to combine disparate kinds of evidence to create hypotheses explaining the major evolutionary transitions leading to higher taxa is the topic of Chapter 9. This is an example of historical sciences, where the subject matter consists of unique, complex past phenomena, and where most of the empirical evidence necessary for a full account is irretrievably missing. As such, the distinction between scientifically valid hypothesis, and plausible but untestable speculation is a matter as much for philosophy as for biology. Certainly there is no easily defined line, and one person's scientific explanation is another person's 'Just-so story' as such scenarios have sometimes been perjoratively labelled. In the present work, I have maintained the view throughout that how and why, when and where a particular higher taxon evolved constitute rational scientific questions because they relate to actual events in the real world. Given this, the principle of scientific explanation of simplicity, or Occam's Razor, holds; namely that the simplest, least ad hoc explanation for the available empirical information to hand represents the preferred explanatory hypothesis, even if the evidential support for that hypothesis is weak in an absolute sense. Moreover, if correctly formulated, directions are implicit within the hypothesis for seeking further corroborating or refuting evidence, or for developing more realistic and useful concepts to interpret the existing evidence.

From this perspective on the nature of explanation in macroevolution, a view of the evolutionary mechanisms responsible for the origin of new higher taxa emerges. Amongst these the two most fundamental elements are the correlated progression means of acquiring derived characters, and the postulated existence of long, multidimensional ridges in the adaptive landscape available for a suitably evolving lineage to track.

CHAPTER 2

The nature of higher taxa

The long-running controversy over the meaning of species (Hausdorf 2011; Shun-Ichiro 2011) and which is the correct concept of what a species is comes down in essence to the fact that the term 'species' is used for two different purposes. On the one hand it is used to denote a unit of evolution in the sense of a population of interbreeding organisms whose gene pool changes over time: hence the various versions of the biological species concept. In this instance, recognition of a species depends on recognition of the interaction between the organisms that constitute it, namely their potential to exchange genes, or act as parts of a single gene pool: 'the species as an individual' (Ghiselin 1974; Minelli 1993). On the other hand, the same term is applied to a rank, preferably the lowest rank, in the hierarchical classification system: hence the various versions of the phenotypic species concept. In this case, recognition of a species depends on recognition of the characters that constitute the definition of one particular species rather than another: 'the species as a class'. Which of these is the 'correct' concept depends on the context. Studies of evolutionary mechanisms by and large require the former; studies of taxonomic diversity the latter.

There is no such conceptual ambiguity about the definition of a supraspecific taxon as a group of phylogenetically related species. Unlike the organisms constituting a species, the species constituting a supraspecific taxon do not interact in the sense of behaving as the functionally related parts of a whole. There are occasional claims that a clade, a monophyletic taxon, has some properties of an individual, namely a 'birth' or origin, a history of change through time, and eventually a 'death' or extinction. It should therefore, the argument goes,

be regarded as an individual entity. However, the significant point is that supraspecific taxa including clades are defined and recognized solely on the basis of the characters shared by their member species, the synapomorphies, and this is a property of a class rather than of an individual consisting of interacting parts.

Amongst the hierarchically arranged ranks of supraspecific taxa—genera, families, orders, classes, etc.—the question of which can be meaningfully described as 'higher' as distinct from 'lower' is less easily answered, but it must be addressed if the study of the origin of higher taxa is to make scientific sense. (One thing that must be stressed is that the taxonomic sense in which 'higher' is used has no relationship whatsoever to the discredited sense of 'higher' and 'lower' levels of being, as epitomized by the *scala naturae* of pre-Darwinian thinking.) There are several helpful ways of illuminating this subjective, though by no means vacuous issue. First, and of course traditionally, a higher taxon is one that occupies a higher level in the Linnean hierarchy of categories: a taxonomic concept. Second, a more graphic way to express this is to describe a higher taxon as one that occupies a sufficiently distinctive region of morphospace: a phenotypic concept. Third, and a correlate if not a consequence of the latter, a higher taxon is one that occupies a distinctive region of an adaptive landscape: an ecological concept. Each of these approaches must be explored.

2.1 The Linnean hierarchy, the phylocode, and higher taxa

The classical Linnean hierarchy of several taxonomic ranks nested within one another—species,

The Origin of Higher Taxa, First Edition. T. S. Kemp.
© T. S. Kemp 2016. Published 2016 by Oxford University Press and The University of Chicago Press.

genera, families, orders, classes, and phyla, plus an assortment of prefixes such as sub, infra, and super for intermediate levels—permits a higher taxon to be defined arbitrarily as one that is assigned a rank deemed to be suitably inclusive. Thus, for example, invertebrate phyla such as Arthropoda or Echinodermata, and also their contained classes like Crustacea and Hexapoda, and Echinoidea and Asteroidea are universally acceptable as higher taxa in this sense, as are the traditional living vertebrate classes such as Cyclostomata, Chondrichthyes, Osteichthyes, Amphibia, Aves, and Mammalia, and probably most of their principal contained subgroups, for example Teleostei, Anura, and Cetacea. Much responsibility lay in the hands of the taxonomists who created these traditional classifications, and the criteria they chose to apply in allocating ranks to particular taxa. In summarizing centuries of taxonomic procedure, Simpson (1961) and Mayr (1969), for example, agreed that all taxa must be monophyletic, although not at that time in the strictly cladistic sense of including all the descendants of an ancestor and which therefore allowed the creation of paraphyletic groups like Bryophyta and Reptilia. When ranking the taxa, the principal criterion was the magnitude of the morphological gaps that separate one taxon from another, although actually estimating the size of these gaps was necessarily a vague exercise. Indeed, the subjective nature of actually deciding what degree of morphological distinctiveness should correspond to what particular rank was always evident and led to irresolvable differences of opinion. For instance, should the marsupial mammals, Marsupialia, be regarded as an order, and therefore equivalent to a single placental order (Gregory 1910; Simpson 1945), or as a cohort equivalent to the whole of Placentalia (McKenna and Bell 1997)?

In the subsequent laudable pursuit of objective criteria for classification, the use of degrees of morphological difference and its associated application to the ranking of clades has been more or less abandoned in modern cladistic methodology. From the latter perspective, the only issue of systematic substance is genealogical relationship based upon the sequence of phylogenetic lineage branching. Cladistic hypotheses are tested by synapomorphy—the sharing of unique characters by putative

sister-groups—but the actual number of synapomorphies supporting a proposed relationship is not of itself of interest other than its bearing on how well that relationship is supported compared to alternative arrangements. Characters unique to one taxon, autapomorphies, are irrelevant because they convey no information that bears on its relationship to any other taxon, and are therefore ignored. Yet it is these very autapomorphies that contribute to the overall morphological differences amongst taxa. Consequently, a cladogram illustrating phylogenetic relationships frequently includes sister-groups that actually hold very different ranks according to the criteria of the Linnean hierarchy (Fig. 2.1a and b). This is particularly frequent in the case of fossils; for example, the fossil fish species *Pholidophorus bechei* is, according to Patterson and Rosen's (1977) early application of rigorous cladistic methodology, the sister-taxon of the entire highly derived and diverse superorder Teleostei. Assignment of rank on the basis solely of synapomorphy while ignoring autapomorphic characters of descendant taxa renders meaningless any traditional sense of 'higher' as opposed to 'lower' taxa.

The most rigorous, and vigorous response to the problem of assigning ranks to taxa has been the rise of the Phylocode, and the continuing attempts to replace the Linnean system altogether by its rules (Laurin and Cantino 2004, 2006). It is predicated on three principles, the first being that taxa are named monophyletic groups with reference to a particular phylogenetic hypothesis. The second principle is to cease defining taxa by the shared derived characters of its members at all: apomorphy-based definitions. Instead, the definitions refer only to the pattern of relationships of the clades in the cladogram. This may be a node-based definition (Fig. 2.1c), in which the named clade consists of the two sister-groups subtended from a node on the cladogram. Or it could be a stem-based definition (Fig. 2.1c), in which the named clade consists of all the taxa more closely related to a named subtaxon than to any other subtaxon outside the clade, so that it includes the stem as well as the crown subtaxa. This latter procedure is particularly suitable for classifying modern taxa as the crown groups, with more basal, fossil relatives as their respective stem groups (Chapter 4, section 4.1). Together, the crown group

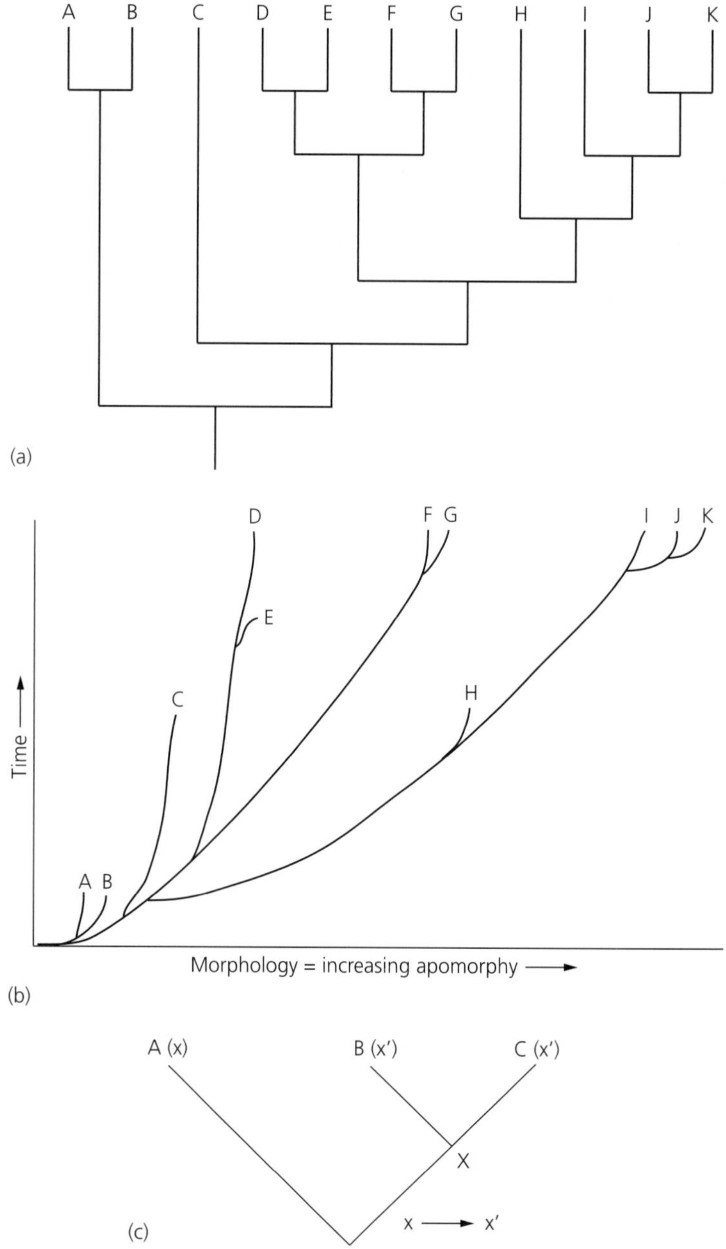

Figure 2.1 (a) A phylogeny expressed as a cladogram based solely on sequence of branching; (b) the same phylogeny but including a dimension of disparity or relative degrees of morphological difference. (c) alternative definitions of monophy: the monophyletic taxon B.C. may be defined as all those taxa possessing derived character x′ (apomorphy-based); all those taxa descended from node X (= node-based); all those taxa more closely related to C than to A (stem-based). Node-based and stem-based taxa can only be recognized after the cladogram has been constructed.

plus the stem group constitute the total group, and the stem group contains all the information held within the fossil record about the origin of the crown group from its last common ancestor with another crown group. The third principle of the phylocode is to abandon all formal ranks for taxa. These are replaced either by simply presenting the cladogram labelled with the taxon names, or by listing the unranked taxa in the order they occur in the cladogram.

Some authors have vigorously opposed the replacement of the traditional Linnean system (e.g. Benton 2000), partly on the grounds that it is too dependent on specific phylogenetic hypotheses, which frequently change in the light of evidence from new specimens and taxonomic revisions, and partly because it rejects so many of the familiar taxa of current standard classifications that non-taxonomic biologists need to use. The principles of the phylocode may well offer an improvement for expressing hypotheses of phylogenetic relationships. However, in the present context its methodology, like that of cladistics generally, excludes information that is relevant to several aspects of the analysis of evolution and diversity, aspects that are quite as important as, and to many minds more interesting than classification per se.

Of no area of evolutionary biology is this more true than in the study of the origin of new higher taxa. Here the focus is on the number and nature of the very autapomorphic characters rejected by cladistic methodology, and the magnitude of the morphological gaps between taxa for which they are a measure. Indeed, if the cladistic interrelationships of the taxa in a particular case are based on a very small set of characters, which they often are, then those relationships are of little relevance to understanding the processes behind major evolutionary transitions. For instance, if two taxa are deemed to be sister-groups on the basis of a few relatively trivial morphological features, or even none at all in some cases using molecular evidence, this implies that their last common ancestor was largely composed of primitive, plesiomorphic characters and that virtually all the significant morphological evolution in the respective descendant taxa occurred subsequently. A good illustration of this point was the discovery of the placental mammalian superorder Afrotheria, which is a triumph of molecular cladistics, and of great biogeographic interest. But a morphological cladistic analysis of Afrotheria has nothing useful to contribute to explaining how the huge morphological disparity arose within a taxon containing animals as different as elephants, sirenians, hyraxes, elephant shrews, tenrecs, golden moles, and the aardvark.

The study of evolution at this level requires recognition of major taxa on the basis of overall morphology, as represented subjectively in a Linnean hierarchical classification. There are plenty of other categories of entities in existence, natural and artefactual, that are continuous with one another and so

require such arbitrary boundaries to be agreed between them, like colours of the rainbow and weight categories of boxers. There is certainly no reason to deny their existence and amenability to research even if the best position for the boundaries may be argued over. Distinction between higher and lower taxa in a multilayered taxonomic hierarchy is another such example.

2.2 Disparity and morphospace: a phenotypic view of higher taxa

A higher taxon viewed as a morphologically defined entity was for a long time associated with the idea of a body plan, a concept originating in pre-evolutionary days and imbued with correspondingly metaphysical overtones, if not an explicitly theological interpretation. Body plans verge on platonic ideals, or archetypes, and the differences found amongst the different species possessing the same body plan were seen as variants of one underlying essential form. Such a way of considering the organisms that belong within a particular higher taxon is wholly inappropriate in the context of transformational evolution because of its implications of immutability and eternal existence. As an analogy for ancestors and evolutionary change respectively, archetypes and variations thereon are misleading. The process of evolution of a higher taxon certainly does not start from an ancestral organism possessing all the attributes of the new body plan: the plan, if that is what it is to be called, has to be assembled over time. Furthermore, divergence into subgroups is not solely a matter of slightly modifying pre-existing attributes, but includes the origin of entirely novel features. To name but one clear example, the molluscan 'archetype' which persisted for a long time in the textbook literature bears little resemblance in fact to bivalves or octopuses. Thus, despite a lingering terminology, the concept of body plans is unhelpful for understanding the nature and origin of higher taxa, and in its place has come the broader idea of disparity.

Any distinction between higher and lower taxa is rendered difficult by the lack of objective measurements of degrees of morphological difference. A number of years ago, Gould (1991) adopted the term 'disparity' to represent the idea of overall morphological differences, and stressed that a means of measuring this property within and between taxa is necessary in order to study a number of aspects of evolution other than simply phylogeny and numbers of taxa. Such issues as, for example, rates of morphological evolution within a clade, degree of conservativeness within different taxa, and the relationship of adaptive radiation to environment all demand estimates of disparity. Since then, several studies have been devoted to methods for quantifying disparity (e.g. Foote 1997; Wills 1998; Ciampaglio et al. 2001; Erwin 2007). In some cases this has consisted of effectively accepting conventional taxonomic rank as a reasonable proxy, despite the subjectivity involved. An order consisting of ten families would be regarded as being more or less twice as disparate as one with only five families. Applying the same logic to the recognition of higher taxa amounts to saying that some particular class or phylum would never have been considered as even a candidate for that rank if it was not morphologically very distinct from any other taxon, which does not advance the argument very much. Morphometric measurements of disparity make use of quantitative, often continuously varying characters, and apply a multitude of statistical techniques to gain a meaningful estimate of differences between and variation within taxa. However, this category of character is much less helpful for distinguishing higher taxa because the question of homology becomes a problem; discrete characters with alternative states are more useful (Foote 1997). Thomas and Reif's (1993) 'skeleton space' is an example in which they defined a set of discrete character states of animal skeletons, such as internal versus external, flexible versus rigid, mode of growth, etc. Although not overall quantitative, and limited in scope to the one functional system, it nevertheless captures a sense of how the morphological differences between different higher taxa can be considered.

Leading on from such multi-character comparisons, the idea of disparity relates to the extremely useful metaphor of multidimensional morphospace. Each character is envisaged as a dimension, and morphospace is the hyper-dimensional space that the values of all the characters together enclose (Fig. 2.2). Any organism occupies one particular point in morphospace depending on the value of its characters, and the distance between points is a measure of overall morphological difference between those organisms. In practice, of course, all the characters could not possibly be considered, but the greater the number that are, the more comprehensively does the distance between points represent overall morphological difference. Considering the modern biota, it is clear on even the most cursory of inspections that morphospace is far from evenly occupied by taxa. Rather, the taxa are clustered into relatively tight discrete groups that are separated by significant gaps from other such groups at the same taxonomic level (e.g. Erwin 2007). This applies at all taxonomic levels, from species within the morphospace occupied by a genus right up to classes within a phylum morphospace. Whether, or to what extent the pattern of clustering is random or deterministic with respect to the environment is an outstanding question. The commonest view is that each cluster represents a group of species that are variants of a common adaptive type of organism (e.g. Wainwright 2007), and so the clustering in morphospace is determined by the existence of clusters of closely similar ecological niches in the habitat. On the other hand it is possible that there is a stochastic basis to an observed pattern of clusters. In principle, stochastic clustering rather than even spacing of taxa could result from a process that in effect throws taxa randomly

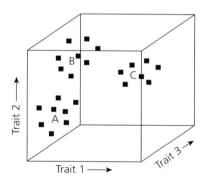

Figure 2.2 Morphological metaphor for higher taxa: taxa clustered together in hypermorphospace. The axes represent values for different phenotypic traits, the blocks taxa at the points indicated by the respective values of their traits, and the letters clusters of taxa representing higher taxa.

into morphospace independently of which points are already occupied. But there is no known mechanism in the context of evolution that could cause this, because every new species necessarily occupies a point in morphospace close to those of its near relatives. Alternatively, a clumping pattern in morphospace with a stochastic element could be generated by evolving lineages following random walks, in which random speciation and random extinction occur without any adaptive considerations at all (Pie and Weitz 2005). However, perhaps the most likely explanation of the pattern of clustering is that it results from a combination of adaptive opportunities, developmental constraints on possible phenotypic evolution, and stochastic processes (Foote 1997; Pie and Weitz 2005; Erwin 2007).

Whatever the cause, this more or less universally observed pattern of disparity helps to clarify the idea of higher taxa and to argue for their reality. A higher taxon is a cluster of species that is separated by a significantly large distance from other such clusters in morphospace. While accepting the arbitrariness of the phrase 'significantly large', two aspects are highlighted by this approach. First, there really are significant-sized gaps between taxa, and

that the greater the gap between taxa the higher those taxa can be considered to be in the taxonomic hierarchy. Second, when considering the origin of higher taxa, the question becomes: how can some diverging lineages of evolving species traverse sufficiently long distances through morphospace to end up being so removed from their ancestors and their contemporaries as to achieve this status of 'higher taxa'?

2.3 The adaptive landscape: an ecological view of higher taxa

The adaptive or fitness landscape is perhaps the most widely applied metaphor in evolutionary biology (Svensson and Calsbeek 2012), and offers an alternative concept to morphospace for describing the nature of higher taxa. In the landscape (Fig. 2.3), the attributes of the organism are represented as the coordinates of a contoured landscape, and the contours themselves represent a measurement of fitness: peaks for maximum and valleys for minimum values. Historically there are two versions of the adaptive landscape. The original one was that

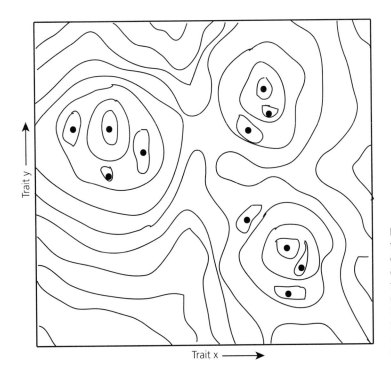

Trait y

Trait x

Figure 2.3 Ecological metaphor for higher taxa: the axes represent trait values; the contours are measures of fitness, the more elevated a point on the landscape, the greater the fitness of a phenotype possessing the coordinate values corresponding to that point; black dots correspond to lower taxa, and higher taxa are clusters of these lower taxa occupying a particular elevated region of an adaptive landscape.

of the population geneticist Sewell Wright (1932), and here the coordinates represent allele frequencies or genotypes. In the later version by Simpson (1944), primarily a palaeontologist, the coordinates represent the values of phenotypic traits. From the Simpsonian perspective, a higher taxon can be thought of as occupying a mountain range of high peaks separated from other such ranges by relatively broad lowlands. The individual peaks within a range are occupied as it were by the subtaxa of the higher taxon, all of which are adapted to relatively similar habitats. For example, there would be thirty-five or so distinct clusters of peaks, one for each of the major animal phyla. As with morphospace, there are no absolute criteria for defining a higher taxon, which in this case means deciding what constitutes a valley of sufficient magnitude and mountain ranges of sufficient altitude to regard the occupiers of the ranges on either side of the valley as representing 'higher' taxa.

Inevitably for such a well-worn metaphor, the adaptive landscape's properties and limitations have been extensively explored and these limitations exposed. For one thing, an adaptive landscape referring to a real-world situation cannot be fixed. Over evolutionary time changing climatic conditions, along with the continual evolution of new taxa and extinction of existing ones, must lead to changes in the optimal combinations of phenotypic characters, and therefore in the positions of the adaptive peaks and valleys (Arnold et al. 2001; Gilchrist and Kingsolver 2001). Indeed, 'seascape' might be a more appropriate expression than 'landscape'. Furthermore, additional features of an adaptive landscape such as transient adaptive 'cols' between adjacent peaks allowing transition from one to the other, and adaptive 'ridges' running long distances across the landscape at a constant elevated height, can be invoked if and when they illustrate specific kinds of relationships between the environment and particular kinds of evolutionary events. The detailed form of the peaks and valleys correspond to different evolutionary scenarios. For example, in an influential contribution, Kauffman (1993) distinguished between a smooth landscape in which there was only one or a few large peaks, and a rugged landscape consisting of a large number of small, relatively similar-sized peaks. In the former

case there is only one or a small number of highly fit phenotypes. In the latter, there are numerous combinations of phenotypic characters that are all more or less equally well adapted. This difference would relate to different kinds of phylogenetic trajectories of lineages traversing the landscape, as discussed in more detail later (see Chapter 6, section 6.1).

All in all, given the broad range of patterns of adaptation in a community that an adaptive landscape can be manipulated to represent, it should not be seen as a hypothesis for predicting evolutionary patterns, but as a graphic means of illustrating how patterns of evolution relate to the environment. In so far as it is the case that a higher taxon consists of sets of species whose distinctive shared morphology reflects adaptations for a shared broad habitat, then the taxon can be described as occupying an upland region, or mountain range, in an adaptive landscape that consists of many such ranges separated by significant lowland areas. Whether this is a realistic picture of the real world needs independent consideration and is discussed in Chapter 6.

To conclude, like morphospace the adaptive landscape metaphor offers helpful ways of thinking about the nature of higher taxa, and it can clarify questions about how they arise. The origin of a higher taxon is thus described as the result of an evolving lineage undertaking a long trek across an adaptive landscape and eventually reaching a new, elevated region. The rate and direction of evolution of any particular such lineage can be related to an appropriately modelled form of adaptive landscape controlling them.

2.4 Molecular taxonomy and higher taxa

As discussed earlier, cladistic methodology in general has little to contribute to the definition of higher taxa because it excludes any estimate of morphological distance between taxa for taxonomic or phylogenetic purposes. Taxa based on molecular sequence evidence is similarly fairly unhelpful, but for a different reason. In this case it is the frequently poor correlation between molecular distance, as indicated by numbers of nucleotide differences between taxa on the one hand and morphological distance, however crudely assessed, on the other. On strictly molecular grounds there might be little justification for accepting some of the otherwise in-disputable higher taxa based on morphology, for example the close molecular similarity indicating a sister-group relationship between whales and hip-pos within the mammalian order Cetartiodactyla. Similarly, as far as molecular evidence is concerned birds are scarcely sufficiently distinct from croco-diles to warrant their status as a separate vertebrate class. Conversely, morphologically quite similar organisms can prove to have very large molecular differences, as for example the tuatara *Sphenodon* compared to a lacertilian like *Iguana*.

As this common lack of close correlation between molecular and morphological distances relates to the highly variable rates of morphological change compared to the generally much lower variability in molecular rates of change, molecular evidence can highlight important and interesting aspects of the process of major evolutionary transitions. While it cannot be used to help define higher taxa or to recognize specific instances of them, it can contrib-ute significantly towards revealing particularly high or low rates of morphological evolution. These differences may in turn relate to environmental con-ditions and thus be important in elucidating pro-cesses by which higher taxa arise.

2.5 The pattern of evolution of higher taxa

Neither the morphospace nor the adaptive land-scape metaphors have a built-in time dimension. Contemporary higher taxa occupy clusters of points separated by gaps in morphospace, or clus-ters of peaks separated by lowlands in an adaptive landscape. However, occupancy of these points or peaks being the result of evolution, the trajectory of an evolutionary lineage over time, through the space or across the landscape, is a visual represen-tation of the origin of new taxa including, where the distance traversed is great enough, higher taxa. This view highlights the question of what drives the lineage. For instance, is its path dictated primar-ily by genetic or phenotypic constraints associated with the pre-existing structure and development of the ancestral phenotype, or is it primarily driven by environmental factors such as the existence of

ecological opportunities for entering new habitats and niches requiring new adaptations? To what ex-tent, if any, is the path a random walk depending on random extinction and origination events? Is the travel roughly constant in rate and direction, or highly variable in these parameters? These are by no means new questions about the causes of evolu-tion and will be returned to in later chapters in the context of particular cases, but meanwhile a num-ber of observations from the fossil record illustrate some general aspects of the pattern by which higher taxa emerge.

Erwin and colleagues (1987) compared the rate of origin of new taxa in the early Palaeozoic with that in the early Mesozoic. Despite commencing from fairly similar, low levels of diversity, the former after the start of the Cambrian and the lat-ter after the end-Permian mass extinction, they found that more higher taxa (phyla, classes, and or-ders) made their first appearance in the earlier time period, while more lower taxa (families and genera) first appeared in the later one. This general, though not entirely universal pattern of higher taxa tend-ing to appear earlier in the fossil record and lower taxa later has since been confirmed several times and for a variety of taxa (e.g. Foote 1997; Ciampa-glio 2004; Valentine 2004; Ruta et al. 2006; Erwin 2007). For example, five or six of the marine bryo-zoan orders appear in the Early Ordovician, despite a very low generic diversity of thirty-three, while only one new order is first recorded in the Mesozoic even though many more genera were in existence by then. The most extreme example of this pattern is the extensively discussed case of the Cambrian explosion (see Chapter 7, section 7.2), in which al-most all the animal phyla are already present by the end of the Cambrian, despite the low generic and ordinal diversity (Gould 1989; Wills et al. 1994). Yet far more lower level taxa of animals evolved during the post-Cambrian Phanerozoic.

Several explanation have been offered for the earlier appearance of higher taxa, and they fall into two broad categories (Holman 1996; Pie and Weitz 2005; Erwin 2007). The first is that the pattern is a result of ecological saturation, in which new broad ecological zones assumed to be necessary for the evolution of radically new kinds of organisms be-come less and less available as the existing number of such zones already occupied increases. Therefore the ecological opportunities still available become increasingly limited to subdivisions of the existing occupied zones, suitable only for relatively similar, and therefore lower level taxa.

The second category of explanation is that the range of possible evolutionary change is limited by constraints internal to the organism. These may be genetic, developmental, or structural constraints, and their effect is to restrict the range of possible new phenotypes that can evolve from an ancestral phenotype. The probability of evolving a radically new kind of organism representing a new higher taxon reduces as more and more of the possible kinds of organisms have already appeared. To put it metaphorically, there are large regions of morphospace that cannot be occupied because such organisms would be inviable; as the morphospace that is potentially available to a radiation becomes filled up, there is less and less opportunity for major evolutionary change to occur and therefore for new higher taxa to evolve.

In addition to these deterministic explanations, other authors have argued that the effect is, at least in part, a secondary artefact of taxonomic structure. This may be due to the earlier, more plesiomorphic members of a higher taxon in the fossil record being less differentiated from their relatives and therefore requiring less morphological distinction to be accepted as members of the new high taxon that, with hindsight, they were. For example, *Archaeopteryx* is recognized as a member of the higher taxon Aves. However, had there been no further evolutionary transition towards modern birds, *Archaeopteryx* would be identified only as a subtaxon of the existing theropod dinosaur taxon Maniraptores (Chapter 8, section 8.2.1). Holman (1996) developed this idea by pointing out that in the history of an evolutionary radiation, the origin of a new higher taxon is also the origin of a new lower taxon, in principle the basalmost species of the new higher taxon. As there are relatively few lower taxa early in the history, a relatively high percentage of these will be basal members of what were to become higher taxa. Later in the history, when there are many lower taxa present, the percentage of these that become higher taxa will be less. Therefore a greater percentage of new higher

relative to new lower taxa will occur earlier than later in the fossil record.

The tendency for the earlier origin of new higher taxa and the later origin of new lower taxa can also be seen in a related phenomenon, which is the apparent decline over time in the rate of morphological evolution in the history of a lineage (Jablonski 2007). Again, this has been observed in most taxa in the fossil record that have been studied, and again the main explanations are either that it is primarily due to declining ecological opportunities, or primarily due to constraints that increasingly limit the possible range of morphological innovation.

In addition to their relatively early origins and higher rates of evolution, another general rule about the appearance of new higher taxa in the fossil record was discovered by Jablonski and Bottjer (1990; Jablonski 2007). They looked at the palaeoenvironmental settings of the first appearances of taxa in the fossil record, and found that higher taxa of benthic, marine groups appeared significantly more frequently in an onshore setting, while lower taxa first appeared more frequently in sediments laid down offshore. This pattern is irrespective of the extent to which the taxa subsequently diversified in one or the other setting. The most obvious difference between the two respective environmental circumstances is that onshore habitats tend to be more fluctuating and unstable in physical and chemical respects. However, why this instability should correlate with the origin of higher taxa is not explained.

2.6 Conclusion: are higher taxa real?

It is scarcely possible to write scientifically about the origin of something unless that something has a real, or at least a theoretically definable existence. If evolving lineages are regarded, as they generally are, as continuous chains of successive species through time generated by speciation events, then the only possible natural discontinuities within the lineage are between species, which are separated by reproductive isolation. If evolving lineages are regarded as the product of anagenetic evolution within a single, continually interbreeding population—a position few authors currently take—there are no natural discontinuities at all. Either way, at least above the species level the whole

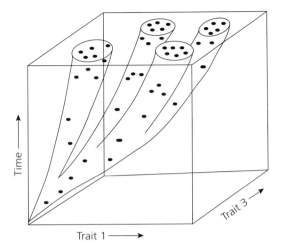

Figure 2.4 Morphological (horizontal) and temporal (vertical) gaps in the record of diverging lineages. The diverging lineages separate both by increasing morphological difference and by extinction of earlier members. The resulting higher taxa are entities separated by boundaries because there is no continuity with either contemporary or common ancestral equivalent taxa.

phylogenetic tree is a single continuum, of which all supraspecific taxa consist of samples taken from the branches. From this perspective, higher taxa could be claimed to have no natural existence because they have no natural boundaries. In practice, however, gaps do exist between clusters of species (Fig. 2.4). There are gaps in a horizontal sense between contemporary species clusters, gaps that arose between lineages in a tree as they diverged from one another. There are also gaps in a vertical sense that arise between different parts of a single lineage due to the extinction of species. These phenotypic gaps are themselves real phenomena because they result from evolutionary processes and not from arbitrary human choice. From this perspective supraspecific taxa are also therefore real entities: they consist of clusters of species separated by these natural gaps. This coincides with the cladistic position that monophyletic taxa are real entities because they are objectively distinguishable, by shared derived characters, from other monophyletic taxa with different derived characters. In this context, the gaps between taxa in a cladogram are represented by the internodes.

However, as discussed earlier in the chapter, cladistic methodology does not allow for the size of the phenotypic gaps between supraspecific taxa to be estimated and represented in a classification, and therefore it cannot discriminate between higher or lower taxa. It is, however, a feature of phylogeny that phenotypic divergence between lineages generally increases over time and so the gaps grow. Gap size is thus a continuum; in principle arbitrary size categories can be established, and then those taxa separated from others by a suitably appropriate sized gap can be defined as higher.

In practice, the magnitude of phenotypic gaps is a subjective matter because it is compounded from too many variable characters to be represented simply by a single index. As with many other gradational or spectral phenomena in biology and elsewhere in the natural world, the significant point is that some quantities or properties change gradually rather than incrementally at the level of investigation they can be subjected to. In the case of supraspecific taxa, these quantities are the amount of phenotypic difference as represented by points in morphospace, and the degree of adaptive distinctness as represented by positions on an adaptive landscape. These points and positions result from evolving lineages pursuing long trajectories through morphospace, and travelling long distances across adaptive landscapes. The processes are real even if the categories of distance travelled that permit the level of a taxon to be agreed on are arbitrary. Few would dispute that each of the following modern groups occupies its own sufficiently distinct region of morphospace to count as a higher taxon: the invertebrate phyla and in many cases their contained classes; the vertebrate taxa of cyclostomes, elasmobranchs, actinopterygians, amphibians, chelonians, lizards, snakes, crocodiles, birds, and mammals; amongst plants, the angiosperms and their constituent families, the gymnosperms, and at least some of the basal spermatophytes. To these must be added innumerable completely extinct but equally distinctive higher taxa known only as fossils.

The conclusion is that higher taxa are indeed real entities and therefore their origin is worthy of investigation.

CHAPTER 3

The nature of organisms

The two essential features of organisms to focus on are: first, that they are highly complex systems consisting of many different parts which structurally and functionally interact in the life of the organism; second, that they evolve, not just by minor adjustments but over time by accumulating major changes in many parts plus the appearance of quite new parts. There is no great mystery about either of these properties in principle, but on the face of it they have a paradoxical relationship to one another. A highly integrated system of different interacting parts tends to be resistant to significant change of any one of the parts because that part would tend to lose its integration within the whole. This idea is encapsulated by a widely accepted tenet of evolutionary theory, namely Fisher's (1930; Orr 2000; Wagner and Zhang 2011) 'cost of complexity' argument that the more complex the organism, the less the probability that a given mutation will prove advantageous. Central to any understanding of the nature of organisms and their evolutionary transitions is explaining how this paradox can be resolved because unquestionably organisms are in fact both complex and evolvable.

Even while acknowledging the integrated nature of organisms, it is impossible in practice simultaneously to take into account all the parts and all the mutual interactions of the parts of any specific organism when attempting to describe it or to infer the evolutionary track through morphospace that brought it about. The methodological solution that must be adopted consists of simplifying the description of the organism to a more tractable level by applying a prior model of the general nature of phenotypes, along with a set of rules assumed to describe how their parts change during evolution. Different models have different degrees and kinds of simplification, and the particular model that is most suitable depends on the nature of the research question addressed as well as the amount of information available. As far as accounting for the evolution of organisms and their characters is concerned, three basic categories of models are in use: atomistic, modular, and correlated. To somewhat overgeneralize, the atomistic model, in which characters are regarded as independent of one another, has been applied mostly in systematics, but it is not an adequate basis for accounting for patterns of character change. The modular model, in which the characters are regarded as clustered together in semi-independent modules, is adequate for explaining relatively short-term evolution but cannot satisfactorily account for major evolutionary transitions. The correlated model, in which all the characters are regarded as potentially and variably integrated with one another, offers the most realistic basis for understanding evolvability at the level of major evolutionary transitions that result in new higher taxa.

These three kinds of models, it must be stressed, do not differ from each other simply in terms of which is true and which are false, for all models are, by their nature and purpose, at best only partial truths. Rather, they differ in which is the most suitable for the actual purpose for which it is being applied. In this respect, phylogenetic and evolutionary biology is no different in principle from most biological research, be it cell biology or ecology or whatever, in all of which arguments can arise that at their core are due to the use of different underlying, but usually unexplicated, simplifying models. Indeed, it can scarcely be over-stressed that behind every study in the field of evolution lurks a

The Origin of Higher Taxa, First Edition. T. S. Kemp.
© T. S. Kemp 2016. Published 2016 by Oxford University Press and The University of Chicago Press.

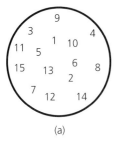 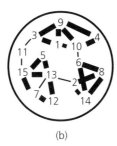 

(a)   (b)   (c)

Figure 3.1 (a) The atomistic model; (b) the modular model; (c) the correlated model. Numbers represent traits, and the thickness of lines represents the relative strength of functional linkages between traits.

model of some kind based on a set of prior assumptions (e.g. Sober 1988, page 237).

3.1 The atomistic model

The simplest model of the organism is to consider it as a set of discrete, mutually independent characters (Fristrup 2001; Rieppel 2001); see Figure 3.1a. As a basis for no more than the description of a single organism or taxon, the only criterion necessary for recognizing a character would be that it can be consistently observed and described. However, virtually all studies in a phylogenetic or evolutionary context involve comparisons of multiple taxa. Therefore an additional criterion is necessary under the atomistic model in order to decide what constitutes the same character in different organisms and can therefore be legitimately compared, namely the criterion of homology. The concept of homology has had a very long history of discussion, from pre-evolutionary days when it equated simply to expressions of the same attribute in different taxa, through to the explicitly evolutionary idea of a character that had evolved in diverging lineages from a common ancestral state (for examples of recent discussions see Ramírez 2007; Scholtz 2010; Vogt et al. 2010). Whatever the fundamental cause of homology is assumed to be, actual examples have to be inferred from the degree of commonality of composition, of position, and where known of embryonic origin, in the different taxa being compared. Virtually all contemporary systematists use cladistics for inferring phylogenetic relationships amongst taxa, a methodology which is predicated on a particular version of the atomistic model. As well as the directly comparative criteria for homology, an important additional one is a measure

of how well a set of putative homologies support one another. Traditionally this is the parsimony, or congruence test (see e.g. Kemp 1999 for an account of the difference between these). It is a means of assessing whether particular characters are homologous by the extent to which their distribution amongst the taxa adds support to the overall most parsimonious, or most congruent cladogram.

In order to move from phylogenetic reconstruction to an explanation of the property of evolvability of organisms, the cladistic version of the atomistic model requires further assumptions about how characters evolve. There was a long post-Darwinian history of attempts to categorize which kinds of characters had higher and which had lower chances of undergoing evolutionary change, which amounted to attributing different 'weights' to different characters. Highly weighted characters were those supposedly unlikely to change, and which were therefore less likely to mislead by convergent evolution in different lineages. Such character weighting was associated with so-called 'evolutionary taxonomy' (Kemp 1985). In due course, an underlying circularity in this reasoning was recognized, on the grounds that there is no test of proposed relative weights, except for inspection of the distribution of the characters on the very phylogenetic tree that the relative weighting had been used to establish in the first place. The scientific unacceptability of this method led to the logical justification of modern morphological cladistics, in which no initial assumptions about the relative probability of evolution of characters is allowed. Instead, the resulting cladistic version of the atomistic model of evolution incorporates the objective assumption that all characters have, *a priori*, an equal probability of transformation during evolution. It may also incorporate the methodological assumption that the probability of a character evolving at any moment in evolutionary time is small, which justifies the parsimony test of a cladogram: the best supported cladogram is the one implying the least number of character transformations, irrespective of what those transformations are (Sober 1988).

The strength of the cladistic version of the atomistic model of evolution lies in its tractability. It is a simple matter to create a matrix of taxa and the states of their presumed homologous characters, and from this to compute the most parsimonious cladogram on the basis of equal weighting of all characters and minimizing homoplasy. The resulting cladogram is amenable to a variety of readily applied statistical tests of how well supported it is both absolutely and relative to other possible cladograms of the same set of taxa, within the constraints of the model. Once a particular cladogram is accepted as the best supported set of taxonomic interrelationships of the taxa under study, it can be read directly as a hypothesis of the phylogenetic interrelationships of the taxa and of the pattern of character evolution amongst them. Since the characters selected from amongst the attributes of the individual organisms are taken to be independent rather than structurally or functionally linked to one another, and since each character is assumed to be equally susceptible to evolutionary change, the cladistic version of the atomistic model has no difficulty at all in accounting for evolvability. Under its terms, any pattern of character change is accepted as possible. At this point it must be stressed that all evolutionary conclusions reached by cladistic analysis are necessarily dependent upon the assumptions of the atomistic model: they are only liable to be correct descriptions of the real evolutionary events to the extent that characters do in fact evolve independently of one another, and if all character transitions could in fact occur with equal probabilities. Clearly neither of these assumptions is true, given the integrated complexity of real organisms.

3.1.1 Limitations of the atomistic model

The weakness of the atomistic model for evolutionary study arises from two related issues. The first concerns the arbitrary nature of boundaries between characters that must be treated methodologically as if they are absolute, discrete entities; the second is ignoring the interdependence amongst structurally and functionally integrated parts of an organism, and the effect this has on their evolution. What is a 'character', and what determines the probability of it evolving from one state to another?

There is a substantial literature stretching back throughout the history of biological systematics discussing what a 'character' is, and what are the appropriate criteria for recognizing one (e.g. contributions in Wagner 2001; Scholtz 2010; Vogt et al. 2010). Since systematics is concerned with formalizing differences between taxa, the tendency has been to define a character as a structural part of an organism that differs in some respects from the equivalent part of other organisms. However, organisms are not divided naturally into discrete characters but consist rather of a hierarchical organization. Any proposed character is simultaneously a part of a higher-level character and itself composed of lower-level parts. There is no objective guidance about which level is appropriate as the usable systematic character. For example, a mammalian dentition might be described as a single character (carnivorous), or each individual tooth described as a separate character (incisor, canine, etc.), or indeed each part of each tooth so described (M1 protocone, M2 parastylar cusp, etc.). The decision on which level of characters to use in a phylogenetic analysis is, on the face of it, entirely arbitrary. Yet the effect is that the dentition may be viewed as just one character or as twenty-one characters, with an obvious effect on the extent to which dental compared to nondental characters contribute to the phylogenetic conclusion. More sophisticated versions of character selection look to embryology and function as well as structure in the search for less subjective criteria for recognizing unit parts of organisms, but the development and the physiology of an organism, like its morphology, are also deeply integrated systems lacking clear natural boundaries between parts. Every practising morphological taxonomist can quote cases where different experts have published different cladograms of a taxon for no other discernable reason than that they chose different characters from the same set of organisms. It is an inevitable consequence of the subjectivity involved in the process of dividing up a complex, integrated entity into discrete parts.

The second weakness of the atomistic model when applied to evolutionary study arises from failure to consider the functional interrelationships amongst the parts. In reality all characters, however defined and recognized, cannot possibly have an equal probability of undergoing an evolutionary change. First, the environment to which the organism as a whole is adapted will dictate that some characters are under directional selection for change while others are under stabilizing selection to remain unchanged. Second, there are fundamental reasons internal to the organism that must affect the relative probabilities of character changes. The functional interrelationships amongst the parts will determine which characters are loosely integrated and so can change, and which ones are so tightly integrated that they cannot change without disrupting the integration of the organism as a whole and so reducing its fitness. Therefore to assume that all characters are equally informative about phylogenetic relationships is unrealistic and must potentially lead to inaccurately inferred phylogenies, and consequently inaccurate hypotheses about patterns of character evolution over time. (For a sample of the many authors who have discussed this idea of 'internal factors' in evolution, see White 1965; Reidl 1977; 1978; Dullemeijer 1980; Schwenk and Wagner 2001.)

An atomistic model performs a good deal better with molecular sequence data, for in this case a single nucleotide can be objectively recognized as a unit character, independent at least structurally from others. Furthermore, the functional interrelationships amongst nucleotides, such as the effect each one has on protein sequence or gene expression, is relatively easy to discover on chemical grounds, independently of phylogenetic relationships. This simplicity of choice of unit characters, plus the sheer amount of available data, accounts for the substantially greater effectiveness of molecular compared to traditional morphological data in resolving phylogenetic relationships.

In contrast, morphological-based cladistic analyses, including cases of fossil taxa where no molecular data are available, raise the question of how far a cladogram is acceptable as an accurate estimate of the true phylogenetic relationships and implied patterns of character transition. Several recent experiences where an extensive molecular data set has completely refuted previous, morphological-based hypotheses of relationships are a salutary warning of the limitations of an atomistic model applied to non-molecular data. The interrelationships of the invertebrate phyla (Fig. 7.2) and of placental mammalian orders (Fig. 3.2) are two such cases where molecular evidence resulted in a highly radical modification of what had been very well-established and widely accepted phylogenies. Where a single morphological-based cladogram is strongly supported by a good number of characters from a variety of different regions and functional systems of the organism, then there is no reason to reject it. To this extent, the atomistic model of character evolution performs satisfactorily as a basis for phylogenetic reconstruction. However, when there is no single, well-supported cladogram, but instead a number of more or less equally weakly supported but contradictory ones, it is unrealistic to accept any one of them as a significantly better estimate of the true relationships. A more realistic model of how the characters evolved is necessary.

Even more importantly for the present work, the highly oversimplified, unrealistic atomistic model fails to account adequately for how and why particular patterns of character change occurred, and therefore what drove certain major evolutionary lineages long distances through morphospace, involving multiple character transformations and leading to the emergence of new higher taxa. Indeed the only imaginable circumstances in which a set of arbitrarily chosen characters might actually evolve with equal probabilities would be no more than a random walk through morphospace, paying scant heed to the environment. To account for evolvability and the origin of higher taxa, a decidedly more realistic model is needed.

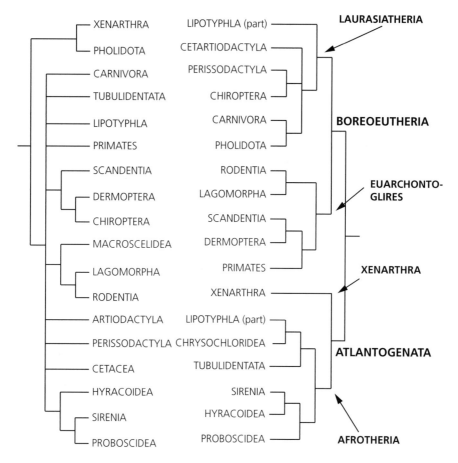

Figure 3.2 Comparison between a traditional morphology-based phylogeny (left) and a current molecular-based phylogeny (right) of placental mammals. Information from (a) Novacek et al. 1988, as figured in Kemp 2005. (b) Murphy et al. 2007. *Genome Research* 17: 413–21.

3.2 The modularity model

Organisms can very conveniently be described as modular in construction; see Fig. 3.1b. Such parts as cells, body segments, limbs, and organs such as hearts and brains, are all sufficiently distinctive and therefore definable morphologically, and also sufficiently integrated structurally in themselves to correspond to any reasonable definition of a module as a 'semi-independent' part. What this means is that the constituent parts of a module are more tightly integrated with each other than they are to the parts of other modules. The result is that a module behaves to a large extent as an independent functional unit in the overall life of the organism and normally, though not necessarily, its constituents develop in

an integrated fashion, and form a spatially continuous entity. It is frequently assumed that modules are also evolutionary units in that they can evolve independently of one another. Thus a limb, for example, might evolve a radical new structure without, say, the heart changing at all.

This traditional, supposedly self-evident view of the organism as modular has been developed in recent years in order to provide the basis of a more realistic model of the organism, a model that offers greater insights into the development and evolution of characters than an atomistic model is capable of providing. Indeed, modularity is currently the most widely accepted model adopted to resolve the integration-versus-evolvability paradox in accounts of the processes underlying phenotypic evolution (Schlosser 2004; Callebaut and Rasskin-Gutman 2005; Wagner et al. 2007).

3.2.1 Evidence for the reality of the modularity model

There are several definitions of modules, but all share the general idea of a high degree of functional interdependence of the parts of a module, coupled with a high degree of independence from other modules (Schlosser 2004; Callebaut and Rasskin-Gutman 2005; Wagner et al. 2007). The classic demonstration of the existence of groups of such interrelated traits was the book *Morphological Integration* by Olson and Miller (1958; Chernoff and Magwene 1999), in which the authors described patterns of significant levels of covariation of traits, such as certain groups of cranial dimensions, within a species population. Maintenance of the integration of such groups of strongly covarying traits in principle could be by means of development processes whereby the genetic mechanism of development simultaneously affects the expression of the separate correlated traits. Alternatively, the maintenance could be by means of natural selection directly favouring the set of trait values that endowed the organism with a higher fitness than would other values.

Since this foundation work, a number of rigorous demonstrations of morphological modules have been undertaken, particularly in mammals. For example, the mouse mandible consists of two regions or modules, an anterior tooth-bearing alveolus region and a posterior region including coronoid, angular, and articular processes. There is a high level of correlation amongst the separate dimensions of each module, but significantly lower correlation between the dimensions of one region compared to those of the other (Atchley and Hall 1991; Cheverud et al. 1997). In a more complex study, Jamniczky and Hallgrímsson (2011) compared intraspecific variation of the base of the skull and part of the cranial circulation system called the circle of Willis, and showed a significant level of covariation between them, indicating a module that consists of more than one histological tissue.

Meanwhile, it had become apparent by the end of the last century that embryological development itself is controlled at the molecular genetic level to a remarkable extent by a developmental system consisting of semi-independent genetic modules (Wagner and Altenberg 1996; Bolker 2000; Segal et al. 2003). More details are presented later (Chapter 5, section 5.3.2), but in the present context, this discovery was taken as indirect encouragement for the view that the phenotype is indeed modular in its fundamental organization, in a way that corresponded somehow to the genetic modularity. A genetic module consists of a set of genes and gene products that act in a coordinated fashion to control aspects of development (see essays in Part 1 of Schlosser and Wagner 2004). The first such genetic module to be recognized was the Hox gene complex, which is responsible for conveying positional information to embryonic cells and structures, such as the whereabouts along the longitudinal axis of the animal a particular kind of segment should develop. Other well-known genetic modules include Hedgehog and its vertebrate homologue Sonic Hedgehog, Wingless, and Notch signalling pathways (see Chapter 5). The discovery that these genetic modules are highly conserved throughout the animal kingdom was surprising enough, and attention moved to the question of how far this genetic modularity is the direct cause of the phenotypic modularity. The possibility was raised by Wagner and Altenberg (1996; Wagner et al. 2007) that what is referred to as the genotype–phenotype map approaches a one-to-one organization, in that a specific genetic module is responsible for a specific phenotype module. It was obvious that there is not a strict one-to-one relationship, but that there is some degree of overlap. A given genetic module may have its major association with a given phenotypic module, but it may also have a minor relationship with other phenotypic modules. Conversely, a phenotypic module is dependent to a small degree on input from several other genetic modules. This is expressed in a much-repeated diagrammatic representation (Fig. 3.3) by Wagner and Altenberg (1996). The term 'variational module' was coined for phenotypic modules that are largely under the control of specific genetic modules. Furthermore, there is a third level of modularity, functional modularity, on the assumption that each phenotypic module is associated with one primary function, although again

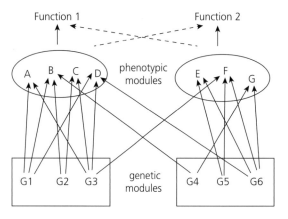

Figure 3.3 Wagner and Altenberg's concept of the mapping of genetic modules onto phenotypic and functional modules. G1–G6 genes organized as two genetic modules and A–G associated with the genes traits organized into two phenotypic modules. There is a near but not complete 1:1 mapping between the two. The phenotypic modules are responsible for two respective functions of the organism, again with near but not complete 1:1 mapping between them. (Redrawn from Wagner and Altenberg 1996.)

overlap is expected, with a phenotypic module having minor associations with additional functional modules.

In fact the relationship between genetic modules and phenotypic modules has proved to be considerably weaker than Wagner and Altenberg's model supposed (Salazar-Ciudad 2009). In the first place, specific genetic modules are almost always involved extensively in the development of a variety of different parts of the organism that cannot by any stretch of the definition be regarded as together constituting an integrated phenotypic or a functional module. For example, the Sonic Hedgehog signalling module in vertebrates is expressed in the development of left–right symmetry, the embryonic formation of neural tube, somites, limb patterning, eyes, hindgut, pancreas, lung, tooth buds, hair follicles, long bone growth, and spermatogenesis (Borycki 2004). None of these can reasonably be regarded as its major single contribution to the organism, with the others minor. The same is true for other well-known modules, and seems to be a general rule. Conversely and equally inimical to the one-to-one concept is that every structural module of the phenotype is invariably under the control of a number of different genetic modules. Even in as simple a case as the size of the beak in Darwin's finches,

the length and the height of the beak respectively are regulated by different genetic modules, despite exhibiting moderate, or in some species quite strong correlation (Wagner et al. 2007). The tetrapod limb is a particularly well-studied example of a putative phenotypic module, yet amongst the signalling pathways expressed during its development, in addition to Sonic Hedgehog just mentioned are various Hox genes and Fgf, Wnt, Tbx, Rng, and numerous others (Tanaka and Tickle 2007). Exactly the same is true in principle of the development of every other adequately studied part of the organism. This constitutes far too extensive a degree of overlap between different genetic, phenotypic, and functional modules for one-to-one mapping between the levels to be a realistic or helpful description of the situation. Therefore, while the concept of modularity of the genetic developmental system remains important, and illuminating for understanding development and evolution (Chapter 5), the idea that this level of modularity is somehow stamped directly onto the phenotype and evolutionary units is not.

If, as now seems clear, the sets of covarying traits of an identifiable phenotypic module do not have a strong direct relationship with a specific molecular genetic module, the question arises of how the integration of these traits is in fact maintained during development and evolution. Where the covariation within empirically determined modules is shown to be under genetic control, it is assumed to involve pleiotropy. This is the phenomenon, well known in classic genetic studies, of a single gene affecting different traits of the phenotype, and therefore of a mutation in a pleiotropic gene having multiple effects. It could be that pleiotropic genes are responsible for the genetic maintenance of integration amongst the traits of a module. Several studies have been aimed at testing whether different parts of a phenotypic module are indeed under the control of pleiotropic genes, with somewhat mixed results. In the first place, studies of the multiple effects of quantitative trait loci and also of gene knockout studies show that pleiotropy is far less common than traditionally believed, with the majority of genes affecting a single trait, and only a few percent of the observed traits affected by multiple genes (Wang et al. 2010; Wagner and Zhang 2011).

Nevertheless, a significant relationship between the targets of pleiotropic genes and the covarying traits has been demonstrated in the mouse mandible (Cheverud et al. 1997), indicating some degree of variational modularity. Even in this case however, there is so much overlap between the modules that Roseman and colleagues (2009) suggested that it is more realistic to consider the traits as distributed through continuous space rather than as discrete modular clusters. In contrast, numerous detailed studies of the range of tissues and organs affected by specific pleiotropic mutations in *Drosophila* and human studies show that their targets are not restricted to sets of traits reasonably regarded as structural or functional modules (Goh et al. 2007).

There are several hypothetical ways in which pleiotropic genes could act to maintain a high level of covariation amongst the traits. The conceptually simplest is that a mutation in a single pleiotropic gene independently affects the developmental pathways of the individual traits of a module in a coordinated way, which is a highly improbable mechanism. A more realistic possibility involves epistatic interactions amongst several genes. Cheverud and colleagues (2004) proposed that the pleiotropic gene does affect the different traits independently, but that the nature of each effect is controlled by other genes. After a mutation in the focal pleiotropic gene, natural selection of the epistatic genes would allow the covariation to be restored over time. They termed this 'differential epistasis' and used it to explain much of the pattern of pleiotropic gene activity that affects different traits in the mouse mandible, as revealed by quantitative trait loci studies. Differential epistasis can also explain why there is a degree of variability in the level of covariation amongst the traits.

A third possible mechanism for pleiotropic gene action is developmental feedback. Klingenberg (2008) discussed the idea that there is direct interaction between the separate developing traits, such that a signal from one modifies the development of others to maintain their mutual integration as a module. A mutation in the pleiotropic gene directly causes a modification in the development of the trait it controls, but it is the development of that trait itself which induces appropriate adjustments to the development of the other traits. Whilst the system is ultimately under genetic control, the mode of signalling between developing traits could be one of many kinds, such as an internal mechanical stimulus, a cellular product, the outcome of interaction with other genes, or an external environmental factor. As an example of this kind of mechanism, Zelditch and colleagues (2008) could explain the empirically observed patterns of integration between the bones, muscles, and teeth in a mouse mandible by what they term epigenetic interactions between the developing muscle tissue and the hard tissues. In another detailed study, Jamniczky and Hallgrímsson (2011) explained their observation of covariation between osteological and vascular anatomy in the skull, referred to earlier, by the known effect that circulatory anatomy has on the development of the surrounding anatomy. Structures and morphological relationships like these examples certainly behave as relatively integrated modules under genetic control. It is not known whether a mutation in a single pleiotropic gene alone can result in the fully coordinated modification, but given the relative paucity and wide range of disparate targets of known pleiotropic genes mentioned, it is highly improbable. A great deal more likely is that it requires the epistatic interactions of many genes, which detracts still further from the idea that the development and evolution of such integrated modules are regulated by single genetic modules. As will become apparent later, this evidence against the existence of a genoptype–phenotype map approximating to a one-to-one relationship is important in the context of explaining multi-character evolution.

3.2.2 The implications of the modularity model for evolution

One of the principal reasons for the interest in a modular model of the organism is that it appears to resolve the paradox that organisms, as highly integrated systems consisting of many interacting parts, should have difficulty changing significantly by evolution. Fisher (1930), as mentioned at the start of this chapter, expressed this as the 'cost of complexity'. He argued that the more complex is the phenotype in terms of numbers of integrated parts, the greater is the probability that a mutation of a given magnitude affecting one part will be disadvantageous because the overall integration of that part in the organism will be reduced. From another perspective, a second theoretical problem for major evolutionary change affecting many integrated traits is referred to as Haldane's 'cost of selection' problem. This refers to Haldane's (1957) argument that there is a limit to how many different traits can be undergoing natural selection simultaneously. By the very definition of selection, a trait undergoing the process requires that more individuals without the trait die than individuals with it, and therefore there must be a number of selective deaths. This is necessarily true of each trait undergoing selection, and yet a population can only support a limited number of deaths without disappearing altogether. Therefore, argued Haldane, there is a limit on the number of traits that can be undergoing selection at the same time. This somewhat controversial idea has been discussed almost exclusively in recent years in the context of the debate over neutral versus selective molecular evolution, but in principle there is no reason why it should not apply to phenotypic characters (Williams 1992; Barton and Partridge 2000). If evolution only concerned single traits or genes, under single natural selection forces there would not be a problem. However, several studies using models of selection investigate the situation when more than one trait is affected by a selection force. The effect on the rate of evolution of a trait that is under directional selection falls when the same environmental pressure is imposing stabilizing selection on a second trait (Wagner 1988; Orr 2000; Griswold 2006). Where multiple traits are under stabilizing selection, or even negative selection, from that selective force, then unsurprisingly the rate of evolution of the focal trait falls even further. Indeed, in principle the point can come where its rate of evolutionary change falls to zero, even though on its own it is still subject to positive selection.

Thus the curious position is reached whereby too high a level of integration of characters runs into Fisher's cost of complexity problem, while too high a level of independence of characters runs into Haldane's cost of selection: both appear to reduce the capacity for adaptive evolution. Obviously something is wrong with this argument, since complex phenotypes presumably can evolve under selection, involving extensive transformation of many characters.

Modularity offers a potential answer to the problem by effectively reducing the number of functionally and structurally interdependent parts in two ways. First, any genetic and developmental mechanisms which ensure that correlated changes occur automatically amongst the separate traits within a module reduce the module from a number of independent variable traits to a single one under selection. If, for example, a mutation in a gene involved in the development of a skeletal element induces by a pleiotropic mechanism integrated modifications to the muscle, nerves, and blood vessels of the limb, then the whole limb can be regarded as equivalent to a single evolving trait. The second way in which modularity can increase evolvability is where, even if a module is composed of independent interacting traits, the number of these will be much smaller than the number of independent traits that would exist if the organism was not modular in construction. The module in its semi-isolation can therefore evolve as if it were a less complex organism in its own right, and so be less subject to the effect of Fisher's cost of complexity. A mutation in any one trait within such a relatively simple entity has a greater probability of increasing fitness while remaining well integrated with its relatively small number of associated traits. Thus, under modularity theory, the modules are assumed to be capable of evolving with no or little reference to other parts of the organism. As long as the separate modules are indeed more or less structurally and functionally independent of one another, then the phenotype as a whole can evolve, separate module by separate module.

There have been a number of empirical studies to test whether organisms do in fact evolve in this way, which is to say whether phenotypic modules revealed by morphometric studies of correlation are evolutionary as well as developmental and structural units. Looking at mammalian skulls, Marroig and colleagues (2009a; 2009b; Porto et al. 2009) analysed the patterns and strengths of covariation in 21 species or genera in 15 orders, on the basis of 3,644 specimens and 34 anatomical landmarks. They found that the pattern of integration of traits amongst the different species, namely the modular structure, is relatively constant. The modules they tested for were oral, nasal, zygomatic, cranial vault, basal, facial, neurocranial, and neurofacial, and they found a generally higher level of covariation within than between these units. However, the actual magnitude of the covariation is surprisingly variable amongst the taxa. This included both covariation within individual modules and covariation averaged over the whole skull. Furthermore, in those species where the average overall covariation is low, the covariation within the modules tends to be high: in other words they are more strongly modular skulls. Conversely, where the overall covariation of the skull traits is high, the covariation amongst the traits within individual modules tends to be low and therefore the modules are less distinct from one another. The probable effect of this difference on evolvability was investigated using a simulated effect of directional selection forces on linked genetically multivariate traits, and it emerged that species with lower overall integration and higher modularity could respond to a wider variety of imposed selection forces (Fig. 3.4). Conversely, a species with a higher overall level of integration is presumably too developmentally or functionally constrained to respond so strongly to most selection forces. The authors' conclusion from this study is that modularity of phenotypic structure does enhance evolvability. Interestingly, the more conservative (i.e. 'primitive') mammals such as didelphids and carnivorans were found to be less modular and so presumed to be less evolvable; the more derived (i.e. 'specialized') taxa such as rodents and primates were more modular and hence supposedly more evolvable. However, there is an alternative interpretation, which is that selection for a more specialized habitat and associated derived morphology generates the higher modularity, and vice versa in the case of selection for conserved habitats and plesiomorphic structure. This latter hypothesis is more consistent with other studies described in the next section that show modules to be readily alterable under selection.

At any event, the theoretical relationship between modularity and evolvability of the phenotype has come to dominate the evolvability debate and the associated issue of the evolution of complexity (Wagner and Altenberg 1996; Schlosser 2002;

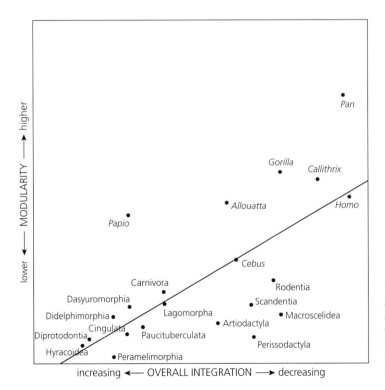

Figure 3.4 The relationship between covariation within modules of the skull and overall covariation amongst the traits of the whole skull. See text for further explanation. (Modified from Marroig et al. 2009b, and reproduced with permission from Springer Science + Business Media.)

Hansen 2003; Wagner et al. 2007; Klingenberg 2008). Indeed, the expression 'evolutionary module' was introduced explicitly for a module able to evolve in relative independence of the rest of the phenotype. However, there are actually several difficulties to accepting modularity as the main mechanism behind major evolutionary change.

3.2.3 The limitations of the modularity model

It should not be forgotten that interpreting the evolving phenotype as modular is adopting a model based on a number of assumptions as well as empirical observations. It is certainly a more realistic model than the atomistic one for explaining evolution involving changes in multiple characters, but nevertheless it does have its own simplifications and limitations. It is demonstrably the case that the mouse mandible or the whole mammalian skull, for example, are constructed of modules in the sense of usefully describable parts. Furthermore, they are also modular in a deeper sense of consisting of measurably covarying traits organized into sets that are to an extent independent of other such sets. However, for such phenotypic modules to behave as evolutionary modules, and therefore to be the major if not sole cause of evolvability, requires at least two more assumptions. The first one is that the majority of the organism's traits that undergo significant evolutionary transformation are organized into such relatively independent modules. The second assumption is that the modules are sufficiently constant in their components over the length of evolutionary distance under consideration. As will be argued, neither of these assumptions is true. On the one hand, many attributes of organisms do not correspond structurally, functionally, or developmentally to modules at all. On the other, the traits that make up a module at any instant in time change over all but relatively very short evolutionary distances, as some traits, as it were, leave and others join a well-integrated group.

There are many constituents of an organism that cannot by any reasonable extension of the concept be described as semi-independent modules, because their structure and function permeate widely throughout the entire organism. As an obvious and familiar example, the vertebrate pituitary gland is a remarkably compact, spatially defined structure that on the face of it should be a prime candidate as a module. However, despite its gross anatomical compactness, it is absurd to regard it as a semi-independent functional unit when its wide range of physiological activities includes such disparate ones as detecting a variety of physiological signals in the blood, and responding by producing a variety of hormones affecting overall growth, sex organs, pigmentation, ionic and osmotic urine composition, lactation, and others. In the process, the pituitary interacts functionally with other anatomical parts of the body such as the hypothalamus and various distant endocrine glands with which it has intimate physiological relationships. Even more obvious cases of manifestly non-modular structure include the circulatory and nervous systems, and various physiological systems concerned with internal regulation of the body that utilize several different organs and processes. Even if described as functional or physiological modules, they nevertheless integrate so closely with so much else of the organism that any idea of semi-independence is misleading. Kemp (2005), as detailed in the next section of this chapter (see Fig. 3.6), discussed mammalian endothermy and how the only description that captures its essence is as a pervasive network of many structures and functions. To consider endothermic physiology as either a module in its own right or as a set of semi-independent modules is not only unrealistic, it also obfuscates rather than clarifies its evolution. In such cases as these, perhaps the modularity model of the phenotype could be rescued by asserting that there are different degrees of semi-independence; some 'modules' are less semi- and some more semi-independent than others, and that the degree of independence of any particular module can change over time. However, this would amount to such a radical modification as to lead towards a different model of the organism altogether, as will be discussed shortly.

The second difficulty in accounting for evolvability by the modularity model concerns the transient nature of modules. A number of studies provide empirical evidence that even well-defined modules alter their components over evolutionary time. Patterns of covariation of traits within a species indicate that anatomical modules are not usually clearly delimited, and overlap one another to varying degrees (Hansen 2003; Zelditch et al. 2008; Roseman et al. 2009). Furthermore, there is even more variation between the traits of a given module when compared amongst related species. In a very detailed study of modularity patterns in the skull of species of *Anolis* lizards, Sanger and colleagues (2011) found that conservation of the pattern of modularity scarcely extends beyond the species level, but rather that it is readily restructured by selection to form different skull shapes for different adaptations. Similarly, Monteiro and colleagues (2005) compared the integration of mandible traits between different species of spiny rats and showed how the traits combined in different correlated groups. A later study of phyllostomid bat jaws showed a similar phenomenon (Monteiro and Nogueira 2009). Here, species with specialized diets such as fruit-eating and blood-sucking have a different pattern of integration amongst the traits, and therefore a different set of mandibular modules, compared to those of less specialized species. This leads to the conclusion that patterns of modularity can be modified and overridden by selection acting independently on specific components. Work on the modularity of the whole mammalian skull shows how the composition of modules changes over evolutionary time at a higher taxonomic level. Based on a large number of anatomical landmarks, Goswami (2006) found six cranial modules in the skull, and measured the degree of correlation of the measurable traits both within and between modules. Although in a majority of species most of the modules were similar, in many cases there was no significant correlation between the traits within a module—in effect they were not modules—which again shows that the modular structure, even in a relatively conservative structure like the mammalian skull, can be broken down, presumably by selection for specific functional attributes like diet. A later study by Goswami and Polly (2010) looked at the relationship between the degree of correlation amongst the parts of a cranial module and the range of variation (the disparity) exhibited by that module amongst a series of carnivore and primate species. They found that modules which are strongly internally correlated within individual species generally showed high disparity when compared amongst different taxa. For most of the weakly correlated modules, there was no significant relationship with disparity at all. This observation supports the conclusion that the integration amongst the traits within a module can be relatively easily weakened as well as the traits themselves shuffled around amongst different modules during the course of the radiation into new species and genera. Presumably these changes are under the control of natural selection, though as suggested in an earlier example, it is not clear whether the degree of modularity affects evolvability, or is itself a consequence of selection for new kinds of habitats.

It is at higher taxonomic levels still that modularity most clearly ceases to offer an adequate explanation for evolutionary change involving multiple characters. Indeed, application of the model can cause actual misinterpretation by its implicit assumption that modules are persistent and not transitory. A simple illustration concerns arthropod body segments. These are a paradigm for modular structure, yet the evolution of tagmatization in different arthropod lineages is not a matter of modifying existing segments that behave as semi-independent and permanent evolutionary modules, but of creating entirely new structures such as the hexapod head or the arachnid cephalothorax. Similarly, the reduction of anatomical segmentation that is found in certain annelids and in the tetrapod vertebrates results in longitudinal muscles that, whatever their evolutionary and developmental origin, have come to have very different functions and functional relationships to other parts of the organism than had the ancestral segmentally arranged muscles. In an example of the positively misleading use of the modularity model, Shubin and David (2004) proposed that the ancestral condition of the tetrapod limb, the sarcopterygian lobed fin (see Chapter 8, section 8.3.1), contained two modules, an endochondral one consisting of the internal bones, and a dermal one consisting of the external finrays. They suggested that change in the relative proportion of the bones to the finrays in different taxa was due to this modular structure. In the case of the tetrapods, the complete loss of finrays and their replacement by endochondral bone digits was, they claimed, an extreme manifestation of the evolvability caused by the modularity. However, this view fails to take any account of the functional relationship between the finrays and endochondral skeleton, both mechanically and physiologically, in their correlated roles as parts of an integrated fin. It is a great deal more plausible to account for the changing proportions of the two parts as being due to natural selection acting on the structure of the appendage as a whole, rather than independently on individual, semi-independent modules. The pertinent question is not how the finrays as an independent entity could be lost without a significant effect on the endochondral part as another independent entity, which is most unlikely, but how the one could reduce and the other enlarge whilst remaining parts of a functioning appendage throughout.

This more realistic approach is illustrated by Young and Hallgrímsson's (2005) investigation of the covariational pattern of traits of the limbs in several mammal species. They measured covariation in size between fore and hind limbs, between serially homologous elements of fore and hind limbs, and between the elements within the limb. There is generally significant covariation within a species both between the fore-limb and hind-limb traits, and amongst the traits within the limbs. However the degree of covariation differs from species to species, which indicates that rather than the limbs consisting of constant modules, the individual parts have the capacity to respond separately to selection by altering their relationships within the limbs. Notably, in taxa with more specialized locomotion such as bats and gibbons, the covariation between limbs is reduced compared to the less specialized quadrupedal species.

A third difficulty with the modularity model, or at least with one of the supposed lines of evidence supporting it, is the lack of a significant one-to-one relationship between identified genetic modules and descriptive phenotypic modules, as already mentioned. This is not critical to the model because other developmental and genetic mechanisms for maintaining modularity can be envisaged, such as pleiotropy and developmental feedback. Nevertheless, the observation does rob the model of one of its underlying justifications, and these alternative processes are equally consistent with the third model of the phenotype, as will become apparent in the next section.

Finally, a rather obvious limitation is that modularity theory has difficulty accounting for the origin of completely novel structures. Broadly speaking, many evolutionary novelties are assembled as new combinations of existing cells and tissues. They cannot be explained simply as modifications of existing modules, as for example the cases of tagmatization and the tetrapod limb mentioned earlier.

To conclude, a complex entity like an organism can be atomized for descriptive and communicative purposes into modules or parts at a single point in its evolutionary history. However ill-defined the boundaries are, both 'horizontally' between parts at the same level and 'vertically' between parts at adjacent levels in a structural hierarchy, it is difficult to see how descriptive modularity can be avoided, or whether it needs to be. But this does not imply that modularity necessarily has a deeper importance as the cause or facilitator of the property of evolvability; whether or not this is the case is a matter of empirical evidence and inference from the nature of organisms (Dassow and Munro 1999). The two main assumptions that must be met if modularity is successfully to account for evolvability are that most if not all of the organism is made up of semi-independent modules, and that the modules are significantly constant in their components through evolutionary time. The evidence from comparative studies of modern organisms is that neither of these assumptions are correct. Furthermore, there is no underlying genetic or developmental system that points uniquely to this role of modularity of structure. It may be that the modularity has a significant role in understanding phenotypic structure and in accounting for evolution at a low taxonomic level. For anything more than short treks through morphospace over brief periods of evolutionary time, however, the modularity model imposes a misleading straitjacket on evolutionary interpretation.

3.3 The correlated progression model

The correlated progression model of the organism redresses the inadequacies of the modular model as an explanation for higher-level evolvability involving patterns of change amongst multiple phenotypic characters associated with multiple functional systems (Kemp 2007a, 2007b).

There is nothing new about the main assumption of the correlated progression model (Fig. 3.1c): in contrast to the modularity model, the phenotype is regarded as a single system of many interacting parts, within which potentially any one part can be functionally integrated with any other. It is assumed that the strength of the functional linkages varies amongst the characters or traits, in the sense that a stronger linkage exists when the precise states of the characters is critical for their effective interaction. A weaker linkage exists when the characters involved can vary to some degree and still interact adequately. In contrast to the modularity model, there is no assumption that groups of strongly linked, correlated traits have lasting relationships and so behave as significantly long-lived evolutionary modules. Rather, the patterns of functional linkages, and therefore the patterns of integration amongst the phenotypic traits, are continually subject to changes over evolutionary time. There are two additional assumptions incorporated into the model. One is that most if not all characters are subject to heritable variation, which is an empirically established fact based on the observation that practically any trait in a species population can be modified by artificial selection. The reason why most traits do not evolve most of the time is a result of stabilizing selection acting on the variation, and not absence of genetic variation in the first place. The second, equally uncontroversial assumption is that natural selection acts on the organism as a whole via its reproductive potential, and that its fitness is an emergent value based on all its interacting character states.

3.3.1 Implications of the correlated progression model for evolution

In an extreme version of this view of phenotypic architecture, all the characters would be supposedly integrated with all the others via a multidimensional network of causes, effects, and correlated actions. Furthermore, if the state of every character were tightly constrained by its functional linkages to all the other characters, no one character would have the capacity to evolve in isolation without a reduction of its integration within the system as a whole, to the detriment of the organism. In other words, Fisher's cost of complexity problem would arise, that as complexity in terms of numbers of integrated characters increases the probability of a successful mutation in any one character decreases. Yet as the model assumes that characters with diverse structure, position, function, and developmental origin have tight functional linkages, neither direct pleiotropic effects by individual gene mutations nor interactions between developmental pathways would be able plausibly to account for the ability of the phenotype to evolve. The correlated progression model addresses this problem of potential lack of evolvability by the assumption that many of the functional linkages between characters are flexible enough to allow a character to vary to a small degree without a significant reduction of its integration in the system; even with many characters interacting, a small change in any single one is still possible. However, there must always be a limit to how much one character can change before it starts to lose its integration with other characters, at which point any further change would start to be deleterious to the system as a whole. No further evolution of that character is now possible unless and until the linked characters have themselves evolved appropriate modifications in the same way (Fig. 3.5). Now the focal character is once again able to change to another small degree. The process is analogous to a line of people walking hand-in-hand forwards, with the rule that no-one is allowed to cease holding their neighbours' hands. At any instant a person may be one step ahead or one step behind their immediate neighbours, but they cannot be more than that without losing hold and thus breaking the line. At no point is anyone more than one step out of line; the line remains intact, yet it nevertheless progresses forwards over time.

In practice, at any instant in evolutionary time, some characters will be more constrained by their functional interrelationships than others, and little if any variation in these is possible. They are subject to stabilizing selection, whereby any variation that does arise will decrease the fitness of the organism and so be selected against. Groups of characters co-constrained in this way will tend to form tightly linked, invariant clusters that have modular-like properties. However, these clusters differ from the modules of modularity theory in a number of

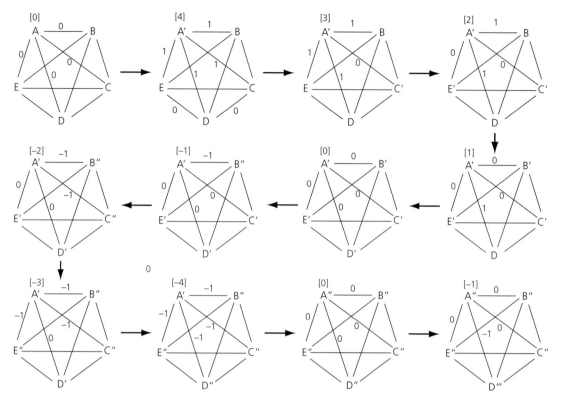

Figure 3.5 The correlated progression model of evolution. The hypothetical organism consists of five traits, A to E, functionally related to one another. The numbers refer to the strength of functional linkages of a trait to the other traits and therefore the inverse of the possibility of a small change in a trait. A small incremental change in a trait is indicated by adding a prime mark: A—A'—A'' etc. (from Kemp 2007b).

respects. First, they can include a wide range of different characters that lack the morphological discreteness and developmental association of modules. Second, they are temporary, transient associations that continually change membership as the strength of the linkages between characters varies over time, and so they cannot behave as evolutionary units. Third, these temporary, module-like clusters do not exist as a consequence of underlying genetic or developmental mechanisms, but by the action of stabilizing natural selection. As the characters evolve and the selection regime changes over evolutionary time, so too will the pattern of functional relationships and strengths of linkages amongst the characters.

The correlated progression model of major evolutionary change has several implications for evolutionary theory (Kemp 2007b).

(a) The pattern of acquisition of new character states, from ancestral to descendant phenotypes along an evolving lineage, consists of small changes in one character at a time, spread over many different characters. These character transformations are functionally related to one another directly or indirectly in the context of the overall adaptation of the evolving organism, hence the term 'correlated progression'. No single character can be seen in isolation as the cause of major evolutionary transitions, so there are no such things as 'key adaptations'. This concept is used to denote a single, newly evolved character that opens up the possibility of invading a new adaptive zone. In fact, evolving a new kind of organism capable of doing this always involves changing many of the integrated characters. The origin of, say, insects required correlated evolution of a great many parts: cuticle, trachea, tagmatization, wings and their neuromuscular apparatus, and so on. No single one of these could have evolved to a significant extent in the absence of evolution of the others. Another common concept that the correlated progression model abandons is pre-adaptation. This is the idea of a specified character that is present in the ancestral habitat and is necessary for invasion of the new one. Because the alleged pre-adaptation is part of the integrated whole ancestor, it is neither more nor less a facilitator of the evolutionary transition than the many other characters with which it is functionally associated.

(b) The precise sequence of small changes in different characters during the course of the correlated progression is more or less random, depending as it does on which of many possible small-effect mutations actually occur. At any instant in evolutionary time there will be numerous characters in which a small change would fit harmoniously within the whole organism. Which one happens next, so to speak, will be unpredictable, but will also have a bearing on exactly what track through morphospace the lineage subsequently takes.

(c) No single character is the focus of natural selection for more than a brief period in evolutionary time because of the limited extent to which it can evolve in isolation of the evolution of correlated characters; the focus will rapidly shift to other characters, particularly those with which the recently modified one happens to be strongly functionally associated. Conversely, no single, simple selection pressure acts for more than a brief time because selection is for the overall fitness of the organism, which is compounded from all its ecological requirements. All the many suggestions made about which single feature of the environment 'drove' the origin of a new kind of organism, be it large and complex metazoans in the Cambrian, tetrapods onto land in the Devonian, or mammals into high-energy lifestyles in the Triassic, are unrealistically over-simple. The lineages were driven, if that is the right word, by the entire complex of

environmental parameters to which their long succession of ancestors were subjected.

(d) The correlated progression model of evolution cannot avoid (Haldane's 1957) dilemma concerning the limit to how many characters can simultaneously be undergoing natural selection, described in the context of modularity in section 3.2.2. This cost of selection argument was widely considered in the context of the neutral theory of molecular evolution, and indeed the discovery of unexpectedly high levels of molecular polymorphism and heterozygosity within species populations was an important part of the original evidence for the theory (King and Jukes 1969; Kimura 1983). In the molecular context, there are a number of proposals for how to avoid the implication, unwelcome to many, that a significant amount of evolutionary change is driven by random drift rather than selection (e.g. Page and Holmes 1998). In the present context of phenotypic evolution involving patterns of changes in many characters over time, the same implication exists, that many of the changes may be selectively neutral or near-neutral. It arises from the assumption that natural selection acts on the fitness of the organism as a whole, to which all the characters contribute, coupled with the observation that most characters are genetically variable. If correct, then logically all the characters are subject to natural selection simultaneously. This is usually ignored by studies that consider a particular focal trait or allele with an assigned selection coefficient as the one under selection, but all the other heritably variable characters potentially affect fitness, and are subject at least to stabilizing if not directional selection. In this situation, cases must be expected where a variation in a character is passively selected because by chance it is a part of the overall fittest phenotype, even if its own contribution to that fitness is zero or perhaps negative. Of course, certain individual character values at certain times will be expected to have a greater contribution than others to overall fitness, and these are the characters that are considered in selection studies. However, this does not detract from the inference that other phenotypic characters are subject to neutral evolution, or to hitch-hiking alongside strongly selected characters based on chance rather than selective value. Not all phenotypic character changes are positively selected and not all phenotypic evolution is driven solely by natural selection. Clearly this inference from the correlated progression model has interesting implications for the directions taken through morphospace by major evolving lineages and adaptive radiations.

3.3.2 Evidence for the correlated progression model

There are several kinds of empirical evidence that support the correlated progression model. The principal argument in its favour is that it is based on a manifestly more comprehensive and realistic view of the structure and biology of organisms than is either the atomistic or the modularity models. For example, Kemp (2006a), in discussing the problem of the origin of endothermy in mammals, gave a systems view of the interrelationships of the structures and processes associated with this functional attribute (Fig. 3.6). There are at least five distinguishable functions of endothemy applying simultaneously: physiologically, thermoregulation ensures constancy of rates of internal reactions which allows for finer regulation of the internal environment; ecologically, thermoregulation extends the range of viable ambient temperature; the high level of sustainable aerobic activity vastly increases endurance such as in high-speed locomotion and extensive foraging; elevated maternal body temperature enhances the rate of juvenile development; and the functioning of a hugely enlarged brain, for which constancy of its temperature and molecular environment is essential, revolutionizes the behavioural repertoire. These functions are all interlinked by their dependence on an elevated basal metabolic rate. However, the high basal metabolic rate itself is dependent upon a formidable array of characters, including mitochondrial size, numbers, and structure, a variety of endocrinal functions at cellular and organ levels, a high-capacity ventilation system for gas exchange, the entire food-acquiring and assimilating equipment, the vascular system that rapidly distributes heat,

oxygen, and glucose around the body, and the activity of the central nervous system regulating the whole system. Consider from first principles how such a system as this could have evolved, where the necessarily high level of integration amongst so many disparate parts means that none could have evolved in isolation, and none could be omitted without loss of the endothermic function. The only feasible way is to build it up by a correlated progression of small, incremental changes of all characters, in no necessarily specified sequence. A small increase in mitochondrial numbers, for example, might slightly increase the metabolic rate with the effect in turn of slightly increasing aerobic activity level, and thermoregulation. However, beyond a small change, any further increase in mitochondria would serve no advantage because the ability of the ventilation system to deliver oxygen, or of the food-capturing ability to acquire adequate fuel, or the circulatory system to deliver them to the muscles would become limiting. Once incremental shifts in these latter have occurred, however, then the point will eventually be reached when there is enough capacity for a further incremental increase in metabolism by the mitochondria to once again be adaptively favoured. In no particularly critical sequence of change of the component parts, the endothermic system as a whole can evolve. The separate functions of endothermy too will all evolve, increment by increment, with the elaboration of the physiological system. It is neither necessary nor realistic to regard one particular function of endothermy as primary or paramount: they can all evolve in concert. This correlated view of mammalian endothermy is supported by the considerable variety of endothermic strategies found in modern mammals. Lovegrove's (2011) detailed comparative review showed that amongst mammalian species many of the individual parameters of endothermy vary, such as basal metabolic rate, maximum aerobic metabolic rate, body temperature, and extent of torpor. This suggests that temperature physiology is perfectly well integrated in all mammals, but that the functional linkages between the parts are sufficiently flexible to permit selection of small changes in individual parts. These differences reflect differences in the detailed ecological niches and habits of different species.

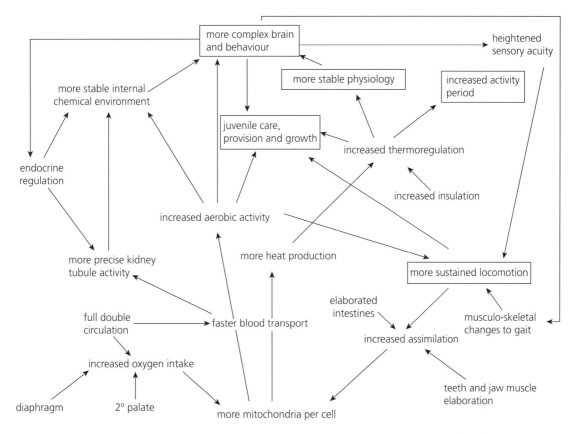

Figure 3.6 Linkages amongst the structures, processes, and functions of mammalian endothermy. The adaptively significant functions are boxed. (From Kemp 2006.)

Budd (1998) gave another example of how correlated progression offers the most realistic explanation for the evolution of complex structure. He considered the nature of arthropod appendages and asked rhetorically which came first, the exoskeleton or the internal musculature, given that neither can function without the other. A correlated evolution of the two, increment by increment of each, coupled also no doubt with appropriate incremental modifications of the central nervous system, the peripheral innervation, the mode of supply of nutrients, etc., is the only plausible explanation. The same logic can be applied to organisms in general, in so far as all are functionally integrated systems in which potentially all the parts interact to produce the biological attributes of the whole.

Computer simulation studies constitute another category of evidence in support of the correlated

progression model. By assuming a system of integrated parts or functions, and applying hypothetical selection forces, the evolutionary behaviour of the system is observed. Niklas (1997, 2004) created a computer simulation of simultaneous selection for several attributes in hypothetical primitive plants (Fig. 3.7). They included mechanical stability, water conservation, light capture, and spore dispersal, which are optimized respectively by different morphological requirements. For example, stability is optimized by a short and compact shape, seed dispersal by tallness, and light capture by outspread branches. He adopted a measure of overall fitness of the plant which represented a combination of the individual fitnesses due to the respective functions, and then allowed the computer to generate random heritable variation in all the characters. In each generation, those that had the highest compromise

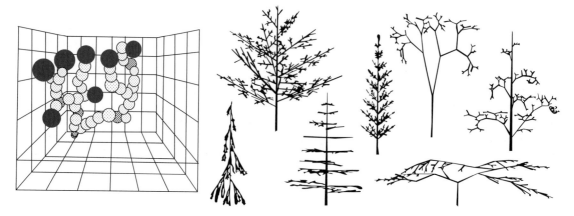

Figure 3.7 Niklas' computer simulation of simultaneous selection for several functions. In this case selection was for a fit compromise between light capture (wide-spreading branches), stability (squat shape), and spore dispersal (high stem). On the left is an example of a computer run. On the right are some of the outcomes, all of high overall fitness but each a different compromise amongst the traits. (From Niklas 1995. Reprinted with permission from the National Academies Press, Copyright 1995, National Academy of Sciences.)

fitness values were selected. The result was an evolutionary trajectory of increasing fitness over the generations that involved changes in all the traits. However, for each separate run of the programme, the actual pattern of change in traits differed, and the final outcome in each case was different. Whilst all the lineages ended up with a high fitness value, each one attained it by a different compromise amongst the traits. In one, the light capturing capacity might be enhanced more at the expense of water conservation, in another the spore dispersal function might be better than the stability, and so on. In a simple way, the experiment illustrates the principle of correlated progression of small changes accumulating in different characters as selection acts on the organism as a whole rather than on either individual or genetically covarying traits. It also illustrates the idea that the actual sequence of character changes is stochastic rather than deterministic, depending on the order in which random mutations occur as well as their effects on overall fitness. It is of considerable interest that many of the plant morphologies resulting from Niklas' programme resemble rather closely a variety of primitive land plants known from the fossil record. This suggests that for all its simplicity, his model is not entirely fanciful.

Marks and Lechowicz (2006a, 2006b; Marks 2007) developed a comparable but more ambitious computer simulation incorporating many more traits. They modelled tree seedlings taking account of 34 different traits, such as root length and distribution, xylem structure, and stomatal behaviour. They noted that each such trait generally contributed to more than one function in the plant, such as acquisition of water, of nitrogen, of light, and of carbon, and therefore that the individual trait values must represent trade-offs between different requirements, as in Niklas' study. To run the simulation, they randomly selected starting values for the traits. Then, generation by generation, they randomly varied the values and at each step selected for a measure of the overall fitness of the seedlings, namely estimated growth rate. Over a sequence of 100 different runs, each involving 2000 generations, the lineages all achieved a high level of fitness, but, as in Niklas' simulation, each one by a different combination of trait values (Marks and Lechowicz 2006b). Even the handful of lineages that managed to achieve an arbitrarily set maximal optimal level showed no significant convergence of trait values. The model was also used to investigate the reaction of a lineage to a change in the environmental parameters, for example an assumed reduction in water availability, with an interesting result (Marks 2007). There was only a weak correlation between the environmental shift and which particular individual traits were affected. In contrast, there was a much higher correlation between the environmental shift and the whole-plant functions that resulted from the combination of a number of single traits.

The comparison between this computer simulation and the assumptions and consequences of the correlated progression model is compelling. First, the selected evolutionary changes in trait values in any one generation were small and the traits affected random. Second, there was no modular structure imposed, either genetic or phenotypic, by pre-setting any constraints over possible combinations of trait values associated with specific whole-plant functions. Presumably temporary patterns of constraint did occur, associated with the trade-offs between traits, because certain combinations did produce an overall reduction in fitness and so were selected against. However, the wide range of different combinations of trait values that eventually evolved indicates that any constrained patterns were transient and continually altering over the generations. Third, selection acting on the entire organism rather than on individual traits nevertheless produces significant change in individual traits. Fourth, the exact pattern of trait changes was unpredictable and with a significant stochastic element to it. The authors were also concerned with the ecological implications of the model, namely that it explains very well why a large variety of species of tree seedlings, all different but apparently equally well adapted, can coexist in a single habitat; indeed they note how very realistic the wide range of simulated optimal tree seedlings is to a natural community. For the present, the significance of this study is that it corroborates the logic of the correlated progression explanation of evolvability, whereby large numbers of characters can transform, yet the phenotype remain an integrated, well-adapted entity throughout.

A quite different but in its own way equally illustrative model is that of Lenski and colleagues (2003) for the evolution of complex characters in general. It is based on self-replicating digital 'organisms', somewhat akin to computer viruses, that have the possibility of acquiring random, spontaneous 'mutations', which in computer terminology are logical functions. The acquired logical functions interact in a way that increases the complexity of the functions available to the organism and this is the property that is selected for, described as computational merit. It is represented as the 'energy' available to the organism as a measure of reproductive fitness. There is a maximum energy availability that could be achieved by as few as 16 mutations, providing they occurred in the right order. However, because the mutations are random it invariably takes many more of them to reach the maximum value. In the course of 50 runs, it was achieved by 26 lineages, but they took between 51 and 721 mutations to arrive there. Furthermore, in the course of evolution the actual sequence of acquisition of specific mutations differed completely from run to run, and no particular mutation was necessary to reach the maximum level. Even deleterious mutations could be fixed because they were part of what was an increase in overall fitness. Despite a relative remoteness from real phenotypic evolution, the analogy offered by this model with the correlated progression model of organism evolution is again obvious. The whole is composed of many interacting parts and the pattern of evolution consists of changes in one part at a time in such a way that the overall integration of the system is maintained, even while its functionality is being optimized.

In another different, formal modelling approach, Walker (2007) considered a model of the phenotype very like the correlated progression model, in which each phenotypic ('morphophysiological') trait affects several functions, some positively and some negatively, and conversely each function is a consequence of multiple traits. He assumed random values for the functional correlations between traits, and for the selection coefficients of traits with respect to each of their roles, positive or negative, and then considered the consequences of the resulting pattern of trade offs and facilitations ('functional architecture') on evolution. The model shows in a formal way how in such a theoretical system different traits evolve at different rates depending on their particular functional relationships with other traits. It also indicates that if the environment changes, and a different selection force arises, the rates of evolution of the traits will also change. As far as it goes, Walker's model supports the general correlated progression view that many traits can be functionally integrated without the need for them to be organized into semi-independent modules for evolution to proceed.

All these simulations of the evolution of a complex organism are artificial in that the values for traits and interrelationships are assumed rather than empirically established. However, the analogy is adequate to demonstrate that such complex systems certainly could evolve in this way, without the need for long-lasting structural or evolutionary modules.

The most compelling evidence that the correlated progression model is a realistic representation of the processes involved in the long treks through high dimensional morphospace that lead to new higher taxa should be the fossil record. If the model is true, then sequences of intermediate-grade fossils would be expected, in each of which relatively small changes in several traits and affecting multiple functions occur. Unfortunately the actual fossil record of the pattern of acquisition of characters leading to higher taxa is very sparse indeed. However, the few cases that are known in adequate detail do support correlated progression, and other more poorly represented examples are certainly compatible with it, as is described at length in later chapters.

3.4 Conclusion

The three kinds of models commonly used in systematic, comparative, and evolutionary biology each has its uses and its limitations. The atomistic model, in which characters are assumed to be independent of one another, underlies a tractable means of classifying and inferring broad outlines of phylogenetic relationships. It cannot, however, capture the complexities of evolutionary change beyond a grossly oversimplified level. The modular model, in which characters are assumed to be combined into semi-independent, covarying groups or modules, that are to a large extent independent of other such modules is an attempt to account for the evolution of complex, integrated organisms. Its weaknesses are that a large part of an organism is not actually modular and that, although demonstrable at the species level, comparative evidence indicates that these modules are too unstable and transient to act as long-term evolutionary elements. The correlated progression model, in which all the characters are assumed to be in principle functionally integrated with each other, offers an alternative explanation of evolvability. Small changes occur in succession in many functionally correlated characters, which leads eventually to major evolutionary transitions while the integration is maintained throughout.

Correlated progression is supported empirically by inference from the systemic nature of modern organisms, and by computer simulations of higher-level evolutionary change under selection. It is therefore the most realistic model for interpreting supraspecific evolution and the origin of higher taxa. The extent to which it is supported by other lines of evidence, palaeontological, developmental, and ecological, will be pursued in the following chapters.

CHAPTER 4

The palaeontological evidence

4.1 Fossils, phylogeny, and ancestry

4.1.1 Incompleteness

Notwithstanding Darwin's own reservations about the incompleteness of the fossil record, for the century or more following the publication of the *Origin of Species* it was generally hoped that the fossil record would eventually reveal at least the outlines of the evolving lineages that led from common ancestors to the members of the living major taxa. Nothing epitomizes better this optimism than the description of *Archaeopteryx* by Thomas Huxley (Huxley 1868, 1886a). Discovered in 1861, it was the perfect moment to find the perfect ancestor, or so it seemed. Less familiar but ultimately even more revealing about the origin of a new higher taxon were the stem mammals, or 'mammal-like reptiles' of South Africa, first described in 1845 (see Owen 1876 for the earliest review of them), which combine ancestral reptilian with derived mammalian characters. Later in the century, the lobe-finned fish *Eusthenopteron* was described by Whiteaves (1883) and duly established as a fossil ancestor of tetrapods. However, examples such as these remained extremely rare, and the vast majority of fossils unearthed had nothing at all to do with the origin of modern major taxa. They either belonged already to modern crown taxa, or else they belonged to equally distinctive groups that had long since disappeared without descendants. Even those that were potentially ancestors invariably turned out to possess characters indicating that they actually terminated short-lived side branches of the main lineages. Indeed, even *Archaeopteryx* suffered this indignity in due course, when a detailed description revealed an unexpectedly specialized jaw articulation and other features of the skull.

Reasons why ancestors obstinately failed to appear as fossils are easy to suggest, once it is appreciated just how incomplete is the fossil record, both stratigraphically and palaeogeographically: many periods of geological time in most geographical areas are not represented by suitably fossiliferous strata. The problem in the particular case of the ancestry of higher taxa is exacerbated by the fact that on average the volume of sedimentary rocks exposed on the earth's surface decreases with age, as a result of losses due to weathering of secondarily exposed sediments, metamorphosis by vulcanism, and subduction of continental margins when tectonic plates collide. As the remote ancestral stages of today's major taxa occurred of the order of hundreds of millions of years ago there are relatively few suitable sedimentary deposits exposed in which they could be found. Even where suitable deposits of the right age and environment do occur, the probability of a member of a species both fossilizing and being discovered is low. Finally, it is eminently probable that early stages in the origin and diversification of major taxa tended to consist of rapidly evolving lineages, existing as small, geographically remote populations, thus reducing yet further the likelihood of discovering them.

As a consequence of these various factors that reduce the probability of finding past organisms, the actual fraction of the total number of extinct species that are preserved, discovered, and described from fossils is minute. An estimate based on comparing the figure of 275,000 or so described Phanerozoic fossil taxa included on the Fossilworks Paleobiology Database (<http://www.paleodatabase.org>) with assumptions about the expected total number of species over this time span suggests that only 1 per cent of all extinct species with hard skeletons

The Origin of Higher Taxa, First Edition. T. S. Kemp.
© T. S. Kemp 2016. Published 2016 by Oxford University Press and The University of Chicago Press.

are known as fossils (Foote 2003). This alone greatly lowers the probability that actual ancestral species are included amongst them.

4.1.2 Phylogeny

On top of the low probability of finding ancestral or intermediate fossil taxa in practice, there was another blow to the idea of using fossils to discover anything about ancestry. During the 1970s and 1980s, the spread of cladistics as the scientifically most objective means of discovering evolutionary relationships led to the logical position that ancestral taxa of any rank, fossil or modern, cannot be recognized as such because they cannot be defined by unique characters. The principle of cladistic methodology is discovering sister-group phylogenetic relationships amongst taxa by identifying shared unique characters: synapomorphies. All groups, or clades, are in principle defined by synapomorphy, and therefore any proposed groups for which there are no synapomorphies are undefinable. Ancestral taxa, including ancestral species, are just such entities, for they have no unique characters, but only a combination of plesiomorphic characters that are also possessed by their own ancestor and derived characters that are also possessed by their descendant.

The corollary that follows from the logic of cladistics is that even if a fossil ancestor were to be found, it could not be recognized as such because of this lack of defining characters. *Archaeopteryx*, for example, could never be shown to be ancestral to birds. Even if the absence of any unique, autapomorphic characters was because *Archaeopteryx* was in fact a direct avian ancestor, it would be assumed that the absence was actually because unique characters had not yet been discovered. It is impossible to know the character state of all of the living *Archaeopteryx*'s characters, so the search can in principle continue indefinitely until one is found, at which point *Archaeopteryx* would become diagnosable as a monophyletic taxon, as did indeed happen in this case and virtually all other cases of fossils at one time or another identified as ancestors. Thus, ancestry itself was relegated to an entirely hypothetical concept that at best is represented by nodes on a cladogram from which sister-taxa diverge, and never represented by real taxa.

This awkward relationship between fossils and ancestry led to the view expressed by several authors that since ancestors are in principle undiscoverable, 'ancestor seeking' in the fossil record is a futile occupation, and furthermore that so few characters are generally preserved in fossils they add little or nothing to cladistic analyses of the vastly more character-rich living members of taxa (e.g. Patterson 1981 and Donaghue et al. 1989, for reviews). Some wanted to exclude fossils from the phylogenetic study of modern taxa altogether. Others were content to just place fossils where they most congruently fitted on pre-existing cladograms of living taxa, like hanging baubles on a Christmas tree, but without allowing them to influence the phylogenetic conclusions (Ax 1987).

A second corollary of cladistics that generated a negative view of the use of the fossil record in phylogenetic analysis was that only characters of the organisms, referred to as intrinsic evidence, are permissible. Extrinsic evidence, consisting of such information as dates of appearances in the fossil record and palaeogeographic provenance, was unacceptable. In large part this was because of the perceived unreliability of such data. Relative age should support hypotheses of the polarity of character changes because the ancestral state must logically pre-date a derived state, but due to incompleteness of the fossil record, the actual appearance of the respective states can be reversed. In comparable fashion, the real dates of the first occurrences of a particular taxon in different areas of the world relate to the areas of origin and the pattern of subsequent vicariant splitting and dispersal of the taxon, but incompleteness of the record can cause spurious apparent dates of earliest occurrence in the different regions. Not only was extrinsic evidence in it own right regarded as inadequate for the purposes of phylogenetic analysis, but it was the character-based cladistic analysis itself that constituted the primary evidence for testing hypotheses about the relative timing of branching events, and of areas of origin and distribution of modern taxa (Fig. 4.1). To incorporate such evidence into the initial cladistic analysis of the modern groups would therefore be tautological.

During this period of its history in the late 1970s and 1980s, palaeontology certainly had a poor reputation amongst many systematic biologists. Those

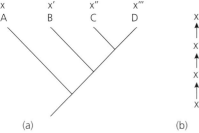
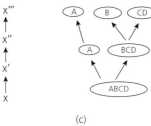

Figure 4.1 (a) A character-based cladogram of four taxa used to test (b) a hypothesis of character polarity, and (c) a historical biogeographical hypothesis for the contemporary distribution of the taxa within three non-contiguous areas.

who supported the anti-fossil view were regarded as ignoring important evidence about evolution, and even as giving comfort to creationists. Those who opposed it and chose to make use of fossils were in their turn regarded as non-scientific story-tellers creating untestable hypotheses. Fortunately the ambiguous attitude to ancestry, whereby it seemed that on the one hand there is no logical reason why ancestors should not turn up in fossil record, but on the other that there is no way of recognizing them if they do, has largely been resolved. An actual ancestor is treated as the hypothetical occupier of a node on the cladogram, but its morphology can be hypothesized on the basis of the characters that define the node (Fig. 4.2). It is assumed that the synapomorphic characters which led to placing the fossil in its position existed in this ancestor. All other characters of the ancestor are inferred to have been in the plesiomorphic condition, as represented by their state at the previous node. Autapomorphic characters of either of the branches derived from the node are of course inferred to have been absent in the ancestor. If it so happens that this reconstructed hypothetical ancestral taxon turns out to be exactly like a known fossil taxon, then the latter can be taken to be a good model for the ancestor, but only as far as its currently known characters are concerned. Smith (1994) referred to such cases as metataxa. The fundamental point is that there is no presumption that any newly discovered characters of the fossil were necessarily present in the hypothetical ancestor. Their status, whether plesiomorphic, synapomorphic, or autapomorphic, has to be tested in the context of the whole cladistic analysis. By this means, the phylogenetic analysis produces a cladogram, the nodes of which represent a sequence of ancestral forms whose reconstructed character

combinations encapsulate everything that the fossil record can reveal about the sequence of evolutionary changes along lineages. This hypothesized pattern of acquisition of the derived characters of a particular higher taxon is what requires explanation if the origin of the taxon is to be understood.

The more general question of how often fossil taxa in practice contribute significantly to phylogenetic reconstruction has been addressed by several studies (for reviews see Donaghue et al. 1989, Kemp 1999, Edgecombe 2010). For example, Gauthier and colleagues (1988) showed that adding fossil evidence to a phylogeny of the main living amniote taxa led to a significant modification of the most parsimonious cladogram, and Edgecombe (2010a) that using modern taxa and fossils together placed the enigmatic tongue worms Pentastomida within Crustacea, but without the fossil evidence, they

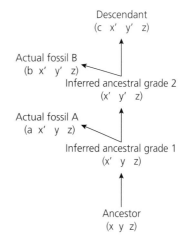

Figure 4.2 The relationship between two fossils, A and B, and a descendant all with their own apomorphies (a, b, and c respectively) and two inferred intermediate stages, ancestral grades 1 and 2, which lack apomorphies.

were resolved as stem arthropods. In their detailed statistical study of a large number of cases, Cobbett and colleagues (2007) also found that fossil taxa significantly affected cladograms, and that there is no evidence that the characters of fossils are any less reliably based than those of modern taxa. Whether fossils can or cannot be useful should no longer be regarded as a matter of principle. In cases where fossils have characters of undisputed homology that form combinations of plesiomorphies and synapomorphies not known in modern taxa, they will contribute significant information to the analysis. Where they do not, they will either fit into existing taxa or remain unresolved at some level, but either way will not affect the cladistic structure. The difficult cases are those where the characters are open to alternative interpretations of homology, in which case one interpretation may affect a cladogram whereas an alternative may not (Donoghue and Purnell 2009). Such cases are a matter of the practicality of recognizing homology which does not differ in principle between fossilized and modern organisms. However, fossil material almost always lacks information on a number of the criteria available in modern organisms for proposing the homology of a structure. Neither embryological nor histological evidence is normally preserved, and spatial landmarks indicating the position of a structure relative to other, soft structures is generally wanting. Several examples of this difficulty will be met later, especially in the context of early invertebrate fossils (Chapter 7).

The greater clarity about how fossil evidence can relate to phylogenetic reconstruction and to illustrating ancestral conditions coincides with the improved precision of terminology described in a previous chapter (Chapter 2, section 2.1). In principle, every node on a cladogram can be named, but it is often more convenient, certainly in the context of discussing ancestry and long-evolving lineages, to distinguish the members of a well-established living taxon from the extinct, more plesiomorphic fossil relatives that represent earlier grades of its lineage (Jefferies 1979; Budd 2001). The living forms constitute the 'crown group', which is defined as the living members, their last common ancestor, and any extinct members descended from that ancestor. The 'stem group' is the paraphyletic, ancestral, or basal taxon that contains all the fossil taxa

that are more closely related to the crown group than they are to its living sister-group. It follows that a member of a crown group possesses all the derived characters defining the living members of that group. A member of a stem group possesses at least one but less than all the characters of a crown group. So, for example, *Archaeopteryx* is a member of the stem group Aves because it is more closely related to modern birds than to their living sister-group, Crocodylia. It is recognized as such because it possesses some characters of living birds, such as feathers, but not all of them, such as loss of teeth and reduced bony tail. Logically, of course, *Tyrannosaurus rex* is also a stem-group bird.

The use of fossil evidence to understand the origin of major new taxa thus becomes a matter of what the stem group of a taxon illustrates about the sequence and pattern in which the derived characters defining the new taxon arose. The greater the number of stem-group members found that differ in their combination of derived and ancestral characters, the more detailed will be the picture of the emergence of the new taxon.

4.1.3 Fossils and molecules

The biomolecular revolution in biology has impacted on the phylogenetics of fossils in two main ways: phylogenetic analysis, and dating of evolutionary events such as divergences between lineages and the starting of adaptive radiations. In both areas, the inexorable increase in DNA sequence data led to prominent conflicts between the fossil and the molecular evidence, and both can have a significant bearing on the origin of higher taxa.

Of the numerous cases of molecular phylogenies conflicting with the purely morphological phylogenies that include fossil evidence, most have been resolved in favour of the molecules. This is primarily due to the virtually limitless amount of molecular data available to resolve increasingly short branches of the cladogram, but also to the relative objectivity of character selection possible for molecules coupled with realistic and computationally tractable models of how molecules evolve. Two prominent examples are the interrelationships of the invertebrate phyla (as discussed in Chapter 7, section 7.1) and those of the placental mammalian orders (Fig. 3.2). In both these cases, overwhelming molecular evidence breaks up several classical taxa, and resolves previously unresolved polytomies. In both, too, the effect has been to alter the nature of inferred ancestral stages of certain major groups. The resulting well-supported phylogenies allow stem-group fossils to be incorporated into the cladogram with greater confidence, leading to interesting new questions about origins and patterns of character evolution along diverging lineages.

There is a serious general implication of this success of molecular evidence. The consequence of demonstrating that morphology can be poor and often quite misleading evidence for discovering relationships of modern taxa where the morphology is potentially completely known, inevitably throws into doubt the accuracy of the necessarily morphological based phylogenies of purely fossil groups, such as trilobites and dinosaurs, where in any case only a very small proportion of the total morphology is known. There is little that can be done about this except to retain a degree of caution, and certainly not to trust too fine a taxonomic resolution. A cladogram of all dinosaurs giving dichotomous branching down to the genus level may be a very convenient classification for communication purposes, but it is quite ill-judged to accept it as likely to be an accurate representation of the true evolutionary tree. The lower the level of the subtaxa, the fewer and the more minor the discriminating characters available, and therefore the greater the uncertainty that they are revealing true relationships.

Molecular data have also revolutionized estimation of dates of divergence between lineages with modern members, with implications for correlating evolutionary events and significant palaeoecological circumstances, and for noting rates of evolution along lineages leading to higher taxa. As with phylogenetics, there have been numerous conflicts between the molecular and the fossil evidence, with the molecular evidence usually indicating an earlier, sometimes very much earlier, divergence date than the time of appearance of the earliest fossil of either of the divergent sister-lineages. Arguments continue about whether this is because of the incompleteness of the fossil record, or the variable rate of molecular evolution; for example, the idea that the clock runs faster at certain times in the history of a clade, such as during the early stages when population size is small or phenotypic evolution rapid has been discussed by Bromham (2003).

Dating a divergence by comparing molecular sequences of modern related forms requires a molecular clock model in order to convert the differences in sequences to evolutionary time. This may be based on an assumed constant rate of evolution, or a statistically based 'relaxed' clock that allows for variation in the rate at different times and in different parts of the phylogeny. However, unlike the molecules versus morphology conflict in phylogeny, in the case of dating lineages the fossils provide their own unique category of information, which is the absolute dates of their existence based on geological evidence extrinsic to the organism and its relationships. Therefore fossil evidence is essential for validating clock models and for independently calibrating a molecular-based time-related phylogeny. Unfortunately, due to the incompleteness of the fossil record, the date of the earliest member of a taxon is not necessarily the date of origin of the taxon. Indeed, it is most unlikely to be. What such a fossil does provide is a latest limit to the estimated divergence date of its lineage, subject to the correctness with which the fossil is classified and the accuracy of its established geochronological age (Parham et al. 2012). However, there is no absolute earliest limit to the date because that depends on how long the lineage was evolving without leaving any fossil record (Benton and Donoghue 2007; Donoghue and Benton 2007). There are various ways to extend backwards in time the estimate of the age of a lineage. One is by a rarefaction analysis, which uses the density of the preserved fossil record of the lineage to calculate the probability that an unpreserved member existed at a specified earlier time. Another is to note the ages of suitable-looking earlier fossil deposits which nevertheless do not contain any members of the lineage. Certain assumptions about the evolutionary process itself may also be made, such as that the diversification of a clade follows a logistic curve. With the necessary assumptions about rate of increase, this allows an estimate of the starting time to be made from the observed rise in diversity over time (Fig. 4.3). However, all such calculations of lower bounds have very obvious limitations.

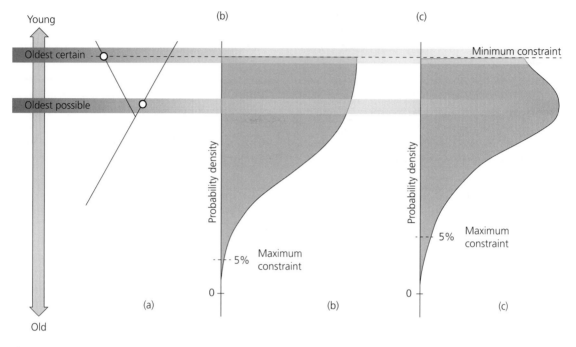

Figure 4.3 Estimating the age of a clade based on an assumed logistic growth model of its diversity. (a) Date of the oldest fossil member of the clade, and of a possible but not certain fossil member. (b) A standard birth–death model of diversification based on the certain member of the clade. (c) A modified model based on the assumption that the uncertain fossil really is a member. (From Donaghue and Benton 2007. Reproduced with permission from Elsevier.)

More recently, the discrepancy between fossil and molecular dates have tended to reduce (see Kumar and Hedges 2009 for many examples of more or less convergent dates). Sometimes this has been due to new or reinterpreted fossils such as the discovery of a Jurassic stem-Placental mammal *Juramaia*, which shifted back the divergence date of marsupial from placental mammals by 35 million years to 160 Ma (Luo et al. 2011). In many others, it has been a consequence of modified models of molecular evolution allowing for rate variation. There does remain a suspicion that the converging dates owe something to the flexibility of the methods available for the molecular analysis of the raw data the better to match the fossil (Donoghue and Benton 2007). Indeed, given the undoubted incompleteness of the fossil record plus the undoubted unreliability of the molecular clock, a synergistic approach combining both sources of information to produce best estimates is necessary.

4.2 Functional anatomy and physiology of fossils

Once the anatomy of a series of hypothetical ancestral stages of a lineage has been reconstructed from the stem-group fossils, the next step in the analysis of the palaeontological evidence is inferring as far as possible their functional anatomy and physiology, in order to assess the adaptive significance of the evolutionary changes that occurred (Kemp 1999). Lacking the possibility of direct experimentation, the preserved anatomy must be interpreted indirectly either by comparison with comparable structure in modern organisms where the function is understood, or by the design method in which function is inferred from the design characteristics of the form on the assumption that morphology reflects the adaptation for its purpose. For example, much has been inferred about the probable size and orientation of musculature from the texture of the surfaces and the gross form of any processes on vertebrate skeletal elements and molluscan shells. The form of teeth, and other kinds of preserved food-collecting organs, are frequently used to infer likely diet.

The spread of computer tomography (CT) scanning of fossils has revolutionized the field of functional anatomy. This includes computerized simulations of locomotory mechanics (e.g. Sellers et al. 2009), in which the limb bones and their joints are scanned, virtually reconstructed, and the possible angles and amplitudes of movement assessed by on-screen manipulation Reconstructions of muscles are then added and the pattern of forces generated and locomotory performance characteristics of the limb estimated.

Another approach of growing importance in biomechanics is the engineering technique of finite element analysis. A 3-D model of a complex shaped structure is built up from CT scanning and then converted to a large number of small elements. A set of simulated forces are applied and the pattern of strain transmitted from them to all parts of the structure is revealed. By applying the method to complex skeletal parts such as vertebrate skulls, and subjecting the models to the forces generated by the reconstructed muscles, much light is thrown on mechanical design. Clear correlations between gross morphology, suture patterns, and strain patterns can lead to hypotheses about the functional design and evolution of the skull (e.g. Jasinoski et al. 2010).

Compared to biomechanics, physiological features are of course a great deal harder to infer from fossil material as they largely depend on soft structures that are not preserved in most fossils. However, techniques of micro-CT and neutron tomography scanning may reveal very fine histological and anatomical structure of skeletal elements that do correlate with physiological functions. For example, bone histology indicates something of the rate of growth of bones (e.g. Chinsamy-Turan 2012a), and the presence or otherwise of zones of arrested growth in tetrapod long bones implies seasonal growth and therefore possibly ectothermic temperature metabolic properties. Laass and colleagues (2010) were able to demonstrate the presence of cartilaginous extensions inside the nasal cavity of the Early Triassic dicynodont synapsid *Lystrosaurus*, comparable to the turbinates of modern mammals which have a thermoregulatory function.

Another relatively newly developed category of techniques now available for functional interpretation is trace element and stable isotope analysis, which can reveal aspects of diet and habitat. For example, the stable oxygen isotope composition of its teeth indicates that the very basal cetacean *Indohyus* matches that of freshwater aquatic vegetation despite its very long limbs suggesting strongly that it occupied a terrestrial habitat (Thewissen et al. 2007). In another example, Botha and colleagues (2005) measured the carbon and oxygen isotope content of the teeth of two common, co-occurring non-mammalian cynodonts of Middle Triassic deposits. One of them, *Diademodon*, shows a low, but significantly variable δO^{18} value and a low δC^{13} value compared to *Cynognathus*. This certainly implies different microhabitats, although it is not easy to interpret confidently. Comparison with the values found in modern crocodiles and mammals suggests that the herbivorous *Diademodon* fed in shadier places closer to water than did the carnivorous *Cynognathus*.

New techniques for investigating functional anatomy and physiology such as these greatly expand the potential for discovering details of the physiological nature and adaptations of extinct taxa, although at present there is no systematic body of knowledge to draw upon. In isolation, much of the functional anatomy of fossil organisms cannot be known in very much detail, and hypotheses suffer from the lack of experimental verification; however, it can be a more significant body of evidence in conjunction with other lines of investigation. To continue the example of the origin of endothermy in mammals (Fig. 3.6), fossil stem mammals in the mid-Permian have several features that can be interpreted as indications of increased levels of activity and metabolic rate. The evolutionary transition coincides with their dispersal to higher latitudes, regions of greater seasonal variation in climate, including times of coolness and times of high aridity. Maintenance of active life under these conditions requires regulation of the animal's internal environment by high energetic processes, for which an elevated metabolic rate is necessary.

4.3 Palaeoenvironmental and palaeoecological reconstruction

One of the focal questions about the origin of new higher taxa is what drives an evolving lineage in the direction it takes through morphospace, and of fundamental relevance must be the ecological conditions of the time period and place of the transition, and the opportunities for new adaptation that they provide. The more that is known about this setting, the more refined can be hypotheses about the nature and extent of environmental control. As there is the caveat that actual ancestral stages are not generally known as fossils, but only as hypothetical constructs, the exact times and places of transitions are also unknowable. Nevertheless, aspects of the broad ecological conditions and significant environmental changes at the estimated times and inferred habitats associated with major evolutionary events can certainly be documented.

Numerous techniques for plotting environmental parameters over Phanerozoic time have been introduced in recent years. Mean global sea temperatures are calculated from oxygen isotope ratios, because water molecules containing the lighter ^{16}O isotope evaporate faster than those containing ^{18}O, an effect that increases with water temperature (Royer et al. 2004). In warmer water, the greater ratio of $^{18}O:^{16}O$ is preserved in the minerals of fossil shells. Over geological time there have been several periods of significantly elevated temperature, greenhouse phases, alternating with reduced temperature, icehouse phases, both with powerful climatic consequences. There are also correlations with a number of prominent evolutionary events. Mayhew and colleagues (2008, 2012) showed a significant correlation between mean global sea-level temperature and biodiversity, and also with rates of origination and extinction. To give a specific example, the maximum placental mammalian diversity was achieved during the elevated temperatures of the Palaeocene and Eocene, a time that saw the origin, or at least the appearance and rapid diversification in the fossil record, of some major new taxa, including whales and bats. Diversity and radiation rapidly declined with the onset of the falling temperatures of the Late Eocene and Oligocene. The catastrophic fall in diversity during several if

not all the great mass extinctions of the past half-billion years, are associated with temperature increases caused by volcanic activity (Twitchett 2006; Sun et al. 2012).

There are several methods for indirectly measuring the atmospheric level of carbon dioxide of the Phanerozoic, one of the main drivers of global temperature (Royer et al. 2004). Plants preferentially use the stable isotope ^{12}C over ^{13}C during photosynthesis. Because soils consist of a combination of both organic and atmospheric sources of carbon, an estimate of atmospheric CO_2 partial pressure is possible from the ratio of the isotopes in fossil soils, given certain assumptions, Other methods are based on the density of stomata in plants, which is inversely related to carbon-dioxide partial pressure, and the ratio of isotopes of boron preserved as a trace element in fossil plankton, which is related to pH, and therefore to the CO_2 concentration in seawater.

Measuring oxygen levels depends on a number of estimates of oxidation activity (Berner et al. 2007; Berner 2009), for example the abundance of reduced carbon and sulphur in sedimentary rocks. The ratio of $^{13}C:^{12}C$, as an indirect measure of photosynthetic activity, and of $^{34}S:^{32}S$, which indicates levels of anaerobic sulphur bacterial levels, can be used. Several important evolutionary events approximately coincide with elevated oxygen levels, such as the Cambrian explosion of the invertebrate phyla, the Devonian conquest of land by invertebrates and vertebrates, and the inferred onset of the evolution of endothermy in stem mammals, all of which are discussed in more detail in Chapter 6.

It is abundantly clear that these various global parameters are interrelated with one another, and in turn intimately related to the patterns of continental drift (Veizer 1988; Goddéris et al. 2012; Nardin et al. 2013), and furthermore that life on earth and its evolution is also an integral part of what has come to be known as the earth systems concept. A central part of the analysis concerns the carbon cycle (Hayes and Waldbauer 2006; Schrag et al. 2013), in which carbon enters rock sediments from both organic and inorganic sources, and some gets buried and often subducted by plate tectonic movements. The carbon is subsequently released into the atmosphere as CO_2 by rock erosion and volcanic output.

Variation in these activities results in variation in the atmospheric CO_2 level, which in turn affects global temperature. These two variables in their turn affect the biosphere in various ways, such as elevated CO_2 increasing plant productivity, which then causes a rise in the O_2 level. As noted, there is evidence that the O_2 level is associated with several major evolutionary events in the history of life.

Earth systems science is some way from a comprehensive model of the interactions between earth, atmosphere, hydrosphere, and biosphere at a global level. Nevertheless, it is unarguable that there is a complex interaction between the biological and the physical domains on earth. Organisms are subjected to the atmospheric and aquatic variables that constitute their environment, but at the same time they contribute to them by their own cycling of oxygen and carbon dioxide. Even while avoiding the somewhat mystical aura of Gaia (Lovelock 1979; Gribbin et al. 2009), it may be assumed that

understanding major evolutionary transitions will have to include an appreciation of the dynamic abiotic setting in which it occurs. A case in point is the great Cambrian explosion, when most of the animal phyla appeared worldwide in the fossil record within a narrow window of time. This involves such a range of geophysical, geochemical, and biological changes that no single, simple cause can explain it, as discussed further in Chapter 6, section 6.5.

The reconstruction of environments on a more local scale is based on a combination of geological methods. Structural sedimentology provides a broad environmental picture such as water depths, coastline features, river flow rates, and land relief. Climatic indicators include identification of fossil plants, pollen, and spores which as a local community are normally climate-specific. Seasonality, precipitation, and wind patterns are often postulated on the basis of climatic models built up from the physical geography.

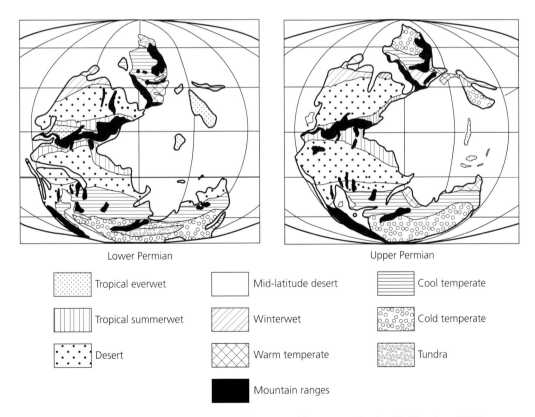

Figure 4.4 The distribution of the continents and biomes during the Lower and the Upper Permian. (Modified from Kemp 2006.)

The exact site of origin of a new taxon not normally being known, neither are the environmental circumstances surrounding it. However is often possible to correlate the inferred functional and physiological biology of a hypothetical ancestor with a particular habitat or biome identified as existing during the critical time period. An example of a comprehensive hypothesis of the global distribution of a series of different terrestrial biomes using a variety of kinds of information is Rees and colleagues' (2002) reconstruction of the Lower and the Middle Permian, on which Kemp (2006) based his attempt to identify the environmental drivers of the origin and early radiation of therapsid stem mammals (Fig. 4.4). In the Lower Permian, desert biomes either side of the tropical zone stretched from west to east Gondwana and so prevented dispersal of terrestrial tetrapods to higher latitudes. However, the palaeoenvironmental evidence points to a partial retreat of the desert barriers in the mid-Permian, opening up a potential dispersal route to higher latitudes. It is hypothesized that the ancestral therapsid biota followed it, and found themselves facing a number of completely new ecological opportunities in the more seasonal and temperate conditions. A very rapid and extensive adaptive radiation into a variety of carnivorous and herbivorous groups followed, as recorded in the fossil record.

Other examples of possible relationships between the origin of new higher taxa and environmental conditions are explored in later chapters.

CHAPTER 5

The developmental evidence

5.1 A brief history

The essential process of evolution is that heritable genetic mutations cause changes in phenotypes by modifying the embryonic or juvenile development. The term 'epigenetic' has usefully been adopted by the 'evo–devo' community to refer to the relationship between the action of the genes and the outcome as an organism (Hallgrímsson and Hall 2011), and is taken to include molecular, cellular, and tissue levels of process. As occurs not infrequently in contemporary biological fields where efforts from different sub-disciplines are involved, terminological differences can arise which, until recognized, can temporarily at least confuse the issue. Thus the term 'epigenesis' is also used in the more restricted sense of environmentally induced modifications to DNA such as methylation of nucleotides, and the effect this has on gene expression (e.g. Petronis 2010; Riddihough and Zahn 2010).

The relationship between development and evolution has figured one way or another throughout the history of systematic and evolutionary theory, and can be traced back to a period before Darwin when embryonic stages of organisms were already being described, compared, and used as additional characters for the systematic arrangement of organisms. Famously, in 1828, von Baer (1828) recognized a parallel between the developmental stages of an organism and the degree of generality of characters amongst taxa in a hierarchical classification. For example, during the development of a mammal, more general or widely distributed vertebrate characters such as vertebrae, cranium, and gill slits tend to appear earlier than do less general, less inclusive, amniote characters such as closure of the gill slits and paired limbs. Exclusively mammalian features such

as separation of the pulmonary and systemic circulations tend to appear even later. In due course, the theory of evolution offered an explanation for this empirical observation: namely that characters of different degrees of taxonomic generality are characteristics of different levels of ancestry: the more general is the character, the more remote is the ancestor that first possessed it, and therefore the more numerous the divergent descendant taxa that inherited it. This strand of thinking continued into the somewhat extreme theory of Haeckel's (1866) that the ontogeny of an organism is a compressed version of its evolutionary history. This, the recapitulation theory, lost credibility once it was realized that embryonic stages do not represent ancestral *adult* stages, but only that earlier embryonic stages of related taxa resemble one another more than do later stages.

Prior to and during the early days of evolutionary thinking, the most important use to which ontogenetic evidence was put concerned the recognition of homologous characters, even if the significance of homology was not at first clearly understood. The paradigm of this is no doubt the discovery by Reichert (1837) that the sound-conducting mammalian ear ossicles are the homologues of the hinge bones of the ancestral reptilian jaw (Fig. 5.1a), but there were numerous other examples where unexpected homologies were discovered in this way, such as the vertebrate thyroid gland with the endostyle of amphioxus, which is a ventral groove in the floor of its pharynx (Fig. 5.1b).

Thus recapitulation theory and modified versions of the embryo-ancestor relationship became a prime source of evidence for discovering phylogenetic relationships amongst taxa, without necessarily inferring anything about evolutionary

The Origin of Higher Taxa, First Edition. T. S. Kemp.
© T. S. Kemp 2016. Published 2016 by Oxford University Press and The University of Chicago Press.

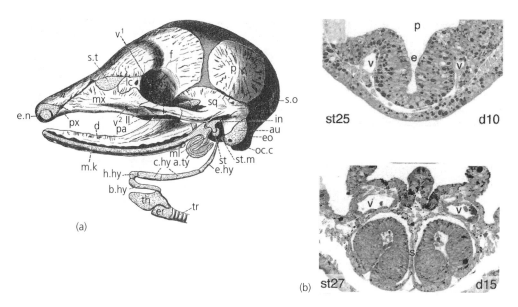

Figure 5.1 Two unexpected homologies revealed by embryology. (a) The incus (in) and malleus (m.l) of the mammalian middle ear occupy the positions of the jaw hinge bones quadrate and articular in the embryo. (b) The thyroid gland of the adult lamprey (below) is derived from an amphioxus-like endostyle groove in the floor of the phartynx of the larval form (above). (a) from Goodrich 1930. (b) from Kluge et al. 2005 reproduced with permission from Springer Science + Business Media.

mechanisms. True, there were discussions about whether 'ontogeny caused phylogeny' or 'phylogeny caused ontogeny', but these were ultimately rather sterile since it was never very clear exactly what either of the expressions actually meant. Meanwhile, the rise of experimental embryology in the twentieth century led to the productive search for the cellular mechanisms underlying morphological development, such as cell movement, induction by one tissue causing another to differentiate, and gradients of diffusible morphogens with variously excitatory and inhibitory effects. In due course a modern reassessment of the relationship between development and evolution commenced, initiated by De Beer's (1958) pioneering work and Gould's (1977) subsequent more detailed clarification The main concern was how modification of the rates of development of different parts of the embryo can result in significant evolutionary changes to the organism. There were also broader issues, such as how selection for embryonic adaptations can affect the evolution of adult characters, and how the

action of novel tissue reactions such as those of the vertebrate neural crest can result in significant reorganization (see Arthur 1997 for a good account). However, for many years virtually nothing was known of the molecular basis for these cellular processes. No postulated morphogen, or cell-signalling molecules were chemically identified, and no explanation was available of how genes and their mutations cause phenotypic characters to develop, let alone evolve. These questions remained entirely within a black box, with Mendelian genes as the input and phenotypic morphology as the output. Even the mid-twentieth-century discovery of the molecular nature of the gene and its relationship to protein structure led to little understanding about the all-important processes of gene function in regulating development.

The implied accusation sometimes levelled at the founders of the synthetic theory of evolution—that they omitted the role of development and therefore generated an incomplete theory—is unfair (e.g. Pigliucci and Müller 2010). No-one could ever deny that development was important, but in the absence of a molecular level of understanding of how it worked, it was quite impossible to incorporate theories of the mechanisms of development into evolutionary theory; the effect would have been more obfuscation than enlightenment. First, the black box had to be opened, and this could not happen until the advent of modern techniques for rapid gene sequencing, for detection of the timing and patterns of specific gene expression, and experimental work on mutations and gene knock-outs in intensively studied model organisms. Commencing around the late twentieth century, this phase of embryological research is now in full swing, under the guise of molecular developmental genetics. The more that is discovered, the greater seems to be the complexity of the networks of interacting genes and gene products that together form the effectors of development. From a plethora of examples and information, a number of general principles have emerged (see e.g. Arthur 1997; Carroll et al. 2001). Two are of particular interest in the context of major evolutionary change and the origin of higher taxa. The first is that much of the genetic developmental system is modular, in the sense that groups of genes and their protein molecule products act together as to some degree independent units that can be deployed at different times and in different parts of the embryo: how might this relate, if at all, to the modular structure of organisms? The second is that the activity of a gene is under the control of what is termed cis-regulation, in which multiple molecules, some with excitatory and some with inhibitory effect, act in concert to dictate the level of transcription activity of that gene: how might this relate to the coordinated changes in multiple genes over evolutionary time?

It may be seen therefore that developmental biology has been and remains an integral part of evolutionary biology in several respects that bear on the question of the origin of higher taxa. First, it continues to provide evidence for inferring aspects of the morphology of ancestral stages and therefore on the pattern of acquisition of derived characters. Second, it offers insight into tissue, cellular, and molecular mechanisms for maintaining phenotypic integrity during major evolutionary transitions. Third, and leading on from the last, by revealing constraints upon and limitations to possible

developmental changes it potentially adds to an understanding of the actual phylogenetic trajectories through morphospace undertaken by lineages. Fourth, by elucidating what kind of effects direct ecological factors can have in modifying development, developmental biology can show whether the process of phenotypic plasticity may play a role in initiating and directing evolutionary pathways across adaptive landscapes.

5.2 Ancestral stages and the pattern of character acquisition inferred from embryos

5.2.1 Recapitulation

At the most elemental, embryonic characters can be used to discover homologies amongst different taxa in which the character has substantially diverged, such as the classic cases mentioned (Fig. 5.1). Embryology has also been applied explicitly in cladistic analysis as a means of determining the polarity of change between two states of a homologous character, in order to distinguish the ancestral, plesiomorphic from the derived, synapomorphic states. This follows from the logic of a von Baerian interpretation of embryonic development, which assumes that successive embryonic stages of an organism illustrate the sequence of states by which the character had evolved from the ancestral state. Use of embryological evidence based on this simplified assumption of the accuracy of the parallel between development and evolution of a descendant organism is analogous to the way in which a temporal series of stem-group fossils could in principle be used to discover an evolutionary sequence of changes in a homologous character. However, just as in the case of fossils, there are a number of epistemological difficulties in using embryonic sequences for this purpose. There is a marked tendency for particular developmental events and even whole stages to be abbreviated, changed in order, lost altogether, or so extensively modified as no longer to appear even to be homologous, presumably in order to increase the efficiency of development (Hall 1995; Rieppel and Kearney 2002). The evolution of specific embryonic, larval, and juvenile adaptations frequently causes divergence of the embryo, larva, or juvenile from the equivalent stage of the ancestral form. Arthur (1997) noted that the ideal von Baerian pattern in which the order of evolution of characters is matched by the order of their development is only well illustrated within the vertebrates. He suggested that this is because the vertebrates tend to provide greater maternal provision and care and therefore there is reduced selection pressure for the developmental stages to evolve their own adaptations for survival. Thus, while embryology may provide significant evidence of phylogenetic relationships based on the possession of shared embryonic features, it sheds scant reliable light on the temporal pattern of evolution of derived characters within a lineage. This is unfortunate, given that there is potentially

100 per cent completeness of observable developmental stages of living organisms, compared to the huge incompleteness of the fossil record of their phenotypic stages of evolution.

Despite these reservations about its reliability, several important hypotheses concerning the evolution of certain higher taxa exist that are primarily based on observational evidence of embryonic sequences, and which relate to the origin of an entirely new kind of organism rather than the history of its individual characters. They mostly involve invertebrate higher taxa, where fossil evidence is sparse. Historically, one of the most influential cases is Haeckel's (1874) blastaea–gastraea theory of the origin of the Bilateria (Fig. 5.2a). He likened a typical blastula stage of development to a hypothetical

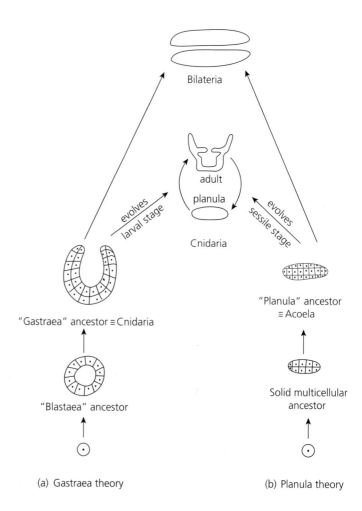

Figure 5.2 Two recapitulationist theories of the origin of bilaterians. (a) Haeckel's gastraea theory assumes that the blastula and gastrula stages in bilaterian development represent successive ancestral stages in their evolution. (b) The planula theory assumes that it is the cnidarian planula larva that represents the common ancestral stage of the bilaterians and cnidarians.

ancestor consisting of a single-layered ball of cells, and inferred that the next stage consisted of a sac-like double layer of cells with a single opening, comparable to an embryonic gastrula. The latter is supposedly represented by the cnidarians. The main rival group of theories about the origin of Bilateria is that the solid, bilateral planula larva of anthozoan Cnidaria (Fig. 5.2b) is comparable to the adult Acoela, which is the simplest living bilaterians. Therefore, goes the argument, the planula represent the common ancestor of both Cnidaria and Bilateria (Hyman 1951; see Minelli 2009), and therefore the early stage of evolution of the Cnidaria is supposedly represented by its modern larval form. This example raises the important point that neither the gastraea nor the planula hypothesis is dependent on the assumption of recapitulation alone, but primarily on the predetermined phylogenetic relationships of the modern taxa. If the superficially gastrula-like cnidarians were not accepted already as close to the ancestry of the Bilateria, or indeed if they did not exist as living organisms at all, then a gastraea theory would never have been taken so seriously. Similarly, if it is argued that the Acoela are the most basal known bilaterians and sister-group to the rest, then a reconstructed hypothetical common ancestor of the two is inferred to have possessed their shared characters, namely to have been a triploblastic, acoelomate, unsegmented organism. It had some degree of antero-posterior body-axis patterning including a central nervous system, and a muscular mesodermal body wall with circular and longitudinal muscles. However, it lacked the anus, cerebral ganglion, and excretory organs of the other bilaterians, and its development was direct, without a larval stage. Continuing the phylogenetic analysis, this hypothetical ancestral bilaterian is found to resemble most closely a planula larva, the two sharing the solid, elongated, bilateral body form.

As is clear from this example, embryological characters may contribute to a phylogenetic analysis and the reconstruction of hypothetical ancestors because they are characters, not because they appear at any particular stage in development. The same is true of other cases of similarities between individual characters of embryos and hypothetical ancestors. The presence of gill pouches in early developmental stages of amniotes implies an aquatic, gilled ancestry, but note that the evidence for this ancestry is overwhelmingly based on a cladistic analysis of known vertebrates in which various gill-bearing fish and amphibian taxa come out as basal to the Amniota. Were amniotes not to possess embryonic gill slits, the hypothesis that they evolved from gilled fishes would be no more affected than it is by the fact that at no stage do developing amniotes possess median fins; it would be more parsimoniously claimed that all signs of gill slits, like dorsal fins, had been suppressed in the course of evolution.

5.2.2 Heterochrony

Heterochrony is the term applied to evolutionary changes in morphology that can be described as modifications in the relative sizes and shapes of features. In fact, many completely new features may be interpreted as extreme cases of heterochrony; for example, the halters of dipteran flies which are highly reduced hind wings, and the mammary glands of mammals which are clusters of modified apocrine sweat glands. There is a plethora of terminology, not always consistent, for different kinds of heterochrony, whether due to accelerations and decelerations of rates of development, or to advances and delays of the start and finish times of processes (Gould 1977; Klingenberg 1998; Gould 2000). There is also ambiguity over whether the very term heterochrony itself is applicable only to the pattern of character changes, or should also include inferences about the developmental causes. Gould (2000) insisted that it should be restricted to the pattern, and pointed out that a heterochronic change in the proportions of a developing embryo need not correspond at all to a heterochronic modification of developmental events. For example, relative reduction in size of a structure might well result from a reduction in the number of cells involved in its differentiation rather than any change in the timing or rate of its growth. Another possibility is an interruption in the extent but not the timing of the expression of a particular gene (Richardson et al. 2009). It is most helpful therefore to consider patterns of heterochronic character changes revealed by comparison of embryonic sequences of related taxa, and what they can suggest about major evolutionary transitions separately from the underlying developmental processes producing them.

Heterochronic evolution is a widespread, commonplace phenomenon, and there is a myriad of examples. Differences in leaf shape, starfish's arm lengths, birds' beak design, and mammalian digit patterns are all potentially describable in heterochronic terms of some parts developing faster or over a longer period of time than other parts, compared to the ancestral condition. Jones and Gould (1999) illustrated a case of shape heterochrony from the fossil record concerning the Lower Jurassic bivalve genus *Gryphaea*, in which the fossilized ontogenetic sequence of the ancestral stage demonstrates a close similarity in shell shape between earlier juvenile stages of the ancestor and adult descendants (Fig. 5.3). At a higher taxonomic level, Weisbecker and colleagues (2008) considered differences between marsupial and placental mammalian limbs and showed that they resulted from different timing of development. The relatively larger size

and precocious development of the fore limbs of marsupials, necessary for the neonate to reach the maternal pouch from the vagina, results from a relative delay in the ossification of the hind limb plus early onset of fore-limb ossification, when compared to placentals. This ubiquity of heterochronic evolutionary change indicates that altering the relative timing and intensity of specific developmental genetic activities is a dominant process involved in generating variation and therefore evolution, with important implications for mechanisms related to the maintenance of phenotypic integrity during transitions. As far as the origin of new higher taxa in particular is concerned, the relevant question is whether heterochrony is only a short-term evolutionary expedient associated with minor change, or whether a heterochronic character change can ever be of sufficient morphological magnitude as to herald the emergence of a new higher taxon. If so, then the process must play a significant role at this level, and there are indeed a number of such cases claimed. The most extreme are hypotheses invoking

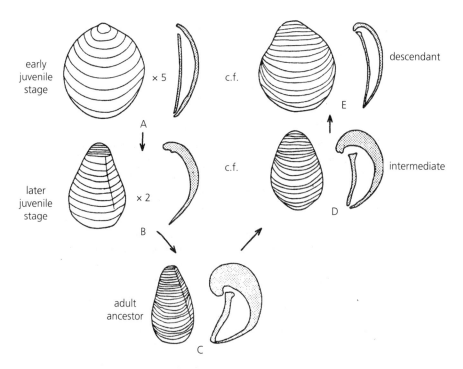

Figure 5.3 Heterochrony in a fossil bivalve. The early and later juvenile stages of the ancestral species on the left are matched morphologically by the two successive descendant evolutionary stages on the right. (From Jones and Gould 1991.)

neoteny, which is the retention of the entire larval form in the sexually mature adults. Returning to the origin of the Bilateria, Nielsen (2008) proposed yet a third hypothesis. He regarded the very simple phylum Placozoa to be the sister-group of the bilaterians, and suggested that it evolved directly from a sponge planula larva by the latter becoming sexually mature and failing to settle and metamorphose (Fig. 5.4a). Garstang's (1894, 1928) theory of the origin of vertebrates from a tadpole larva of

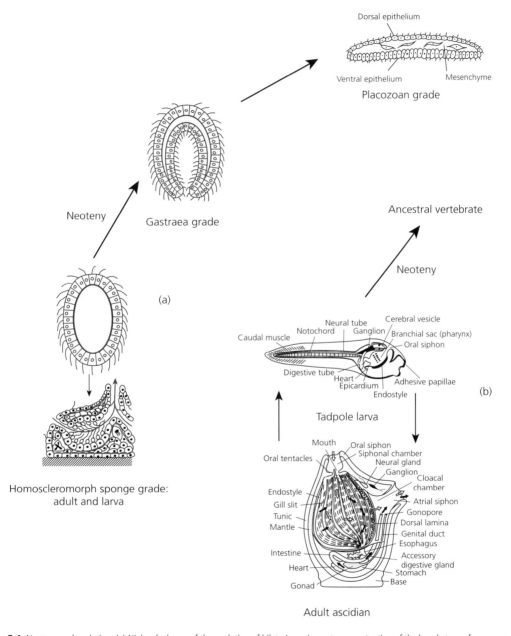

Figure 5.4 Neoteny and evolution. (a) Nielsen's theory of the evolution of bilaterians via neotenous retention of the larval stage of a sponge. (b) Garstang's theory of the origin of vertebrates via neotenous retention of the tadpole larva of an ascidian. ((a) from Nielsen 2008. (b) from Brusca and Brusca 2003.)

an ascidian remains the most celebrated example of the possible origin of a major new taxon by hetero-chrony (Fig. 5.4b). It is based on the musclulariza-tion, notochord, and dorsal nerve cord characters of the larval tail which are comparable in general form to the structures found in the vertebrates. How-ever, as in the case of the Bilateria, again there is an alternative view, namely that the ascidians also evolved from this free-swimming, bilateral ances-tor, by subsequently adding on metamorphosis to a sessile form.

Heterochrony has been implicated in the origin of the minute, meiofaunal phylum Loricifera (Fig. 5.5). Molecular evidence supports a sister-group relation-ship with the Nematomorpha, but the morphological characters of nematomorphs that support this grouping are found not in adult but larval nem-atomorphs (Sørensen et al. 2008). They include the presence of chitin, which is lost in adult nemato-morphs, hexaradial symmetry of the mouthparts, a buccal canal described as twisted, and a diaphragm between thorax and abdomen. The inference is that the common ancestor had a nematomorph-like life cycle with larva and vermiform adult and that the loriciferans lost the adult stage by neoteny (Fig. 5.5).

Rozhov (2010) believes that the origin of the nu-merous echinoderm classes in the early Palaeozoic can be accounted for by heterochronic changes in the relative sizes and combinations of characters of the common ancestral echinoderm. Although he

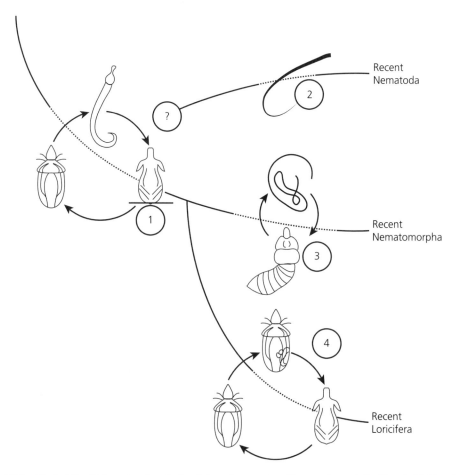

Figure 5.5 Origin of the Loricifera (4) by neoteny from the larval stage of an ancestor with a nematomorph-like life cycle (1). (From Sorensen et al. 2008 reproduced with permission from John Wiley & Sons.)

does not elaborate in any detail, he does make the interesting suggestion that selection for increased dispersal in an essentially benthic group of organisms led to increased length of the larval life by delaying development, and that this in turn facilitated heterochronic variation.

A number of major transitions within the vertebrates may have involved heterochronic modifications in particular important characters. For example, juvenile non-avian theropod dinosaurs have relatively longer fore limbs than do the adults, and therefore a relative limb size more comparable to that of the basal bird *Archaeopteryx*. This observation led Long and McNamara (1995) to conclude that a heterochronic increase in fore-limb size was a significant process in the origin of birds and bird flight. The celebrated theory of the origin of the human skull by neoteny is based on the greater similarity in form of the face and cranial cavity of juvenile chimpanzee and adult human skulls than that between two adults skulls.

5.2.3 Heterotopy

Another mode of evolution in which integrated change in complex parts of the phenotype occurs is termed heterotopy. It consists of a shift in the spatial relationships amongst the parts of the adult compared to the ancestor, and results from a presumed change in the positions in the embryo where specific developmental events occur (Zelditch and Fink 1996). Straightforward examples include the shift of the pelvic fins to a position anterior to the pectoral fins in acanthopterygian teleosts. However, as with heterochrony, it is as well not to assume that a heterotopic modification to the adult anatomy always corresponds in a simple way to a geometrical shift in the deployment of a developmental process to a new place. It may be the case, for example, that the heterotopic change in morphology resulted from subsequent differential growth of tissues after the formation of the structure in its original position.

Zelditch and Fink (1996), in viewing these patterns of morphological change from a wider perspective, pointed out that the distinction between heterochrony and heterotopy is rarely clear. A change in the timing of a developmental event is necessarily associated with a different pattern of embryonic tissues in which it occurs, while a change in the spatial pattern of the developmental events is virtually certain to affect their relative timing. Thus, for example, evolution of changes in arthropod segmentation patterning involves both timing and spatial shifts in the action of Hox genes.

Notwithstanding this cautionary note, there are examples of significant evolutionary transitions associated with the origin of new higher taxa that have been attributed to heterotopy. The evolution of the turtle body plan includes the extraordinary and unique rearrangement of the ribs and limb girdles, so that morphologically the latter come to lie internal to the former (Fig. 5.6). Indeed, this transition has been regarded by some as a serious candidate for a macromutation producing a hopeful monster (Rieppel 2001a). However, the position of the turtle shoulder girdle is less bizarre when compared to basal amniotes rather than to mammals and birds; in the former, the shoulder girdle lies morphologically more anterior to than lateral to the dorsal ribcage. The developmental changes required to achieve the turtle morphology can be seen as a heterotopic shift in the position of the ribs relative to the body wall (Nagashima et al. 2009; Lyson and Joyce 2012). The middle of the carapace is formed in association with the neural arches of the dorsal vertebrae, and the main lateral parts with the dorsal ribs. In the embryo of a typical hard-shelled turtle, a fold develops along the sides, initially between the front- and hind-limb buds, but later being completed as a circle by anterior and posterior extensions (Fig. 5.6). Referred to as the carapacial ridge, the fold consists of a thickening of ectoderm plus underlying mesenchyme. As the ribs of the embryo grow outwards, they appear to be attracted to the carapacial ridge and so remain dorso-lateral, rather than extending ventrally in the body wall internal to the shoulder girdle as in other tetrapods. The major shoulder girdle muscles such as the anterior serratus, retain their ancestral connections with the skeletal elements, which results in a radical change in their topology that is consistent with the modified skeletal anatomy (Cebra-Thomas et al. 2005; Nagashima et al. 2009). However, the muscles associated with fore-limb movements have changed their pattern of attachment to the skeleton in order to maintain their function. The latissimus dorsi, for

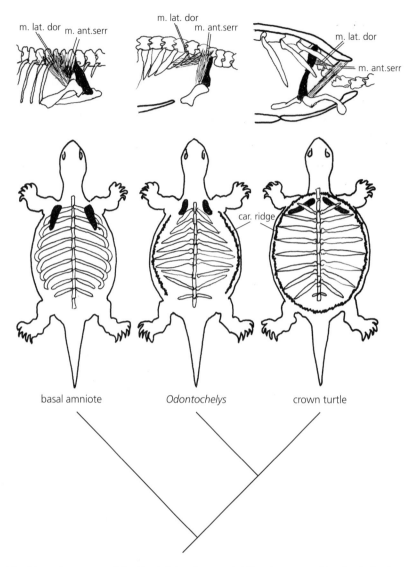

Figure 5.6 The origin of the turtle carapace and shoulder girdle by a heterotopic shift in the position of the ribs. The fossil stem turtle *Odontochelys* is reconstructed as intermediate in this process. car.ridge = carapacial ridge; m.lat.dor = latissimus dorsi muscle; m.ant.serr = anterior serratus muscle; shoulder girdle = solid black. (Modified from Nagashima et al. 2009.)

example, which in basal amniotes grows out from the embryonic limb bud in a postero-dorsal direction over the lateral fascia of the ribcage, in turtles extends forwards to attach to the nuchal plate, the anterior-most element of the carapace and functionally an appropriate new site of attachment. This example illustrates particularly well the principle that a heterotopic change, even as radical as this, includes adjustments that allow retention of the

functional integration of the various parts of the structure involved.

5.2.4 Allometry, miniaturization, and gigantism

When the overall size of an organism evolves and all the individual dimensions change at the same rate, the pattern of change is described as isometric, and the shape remains constant. The situation where different dimensions change at different rates is referred to as allometry, and is a more or less universal phenomenon in evolving lineages that experience significant size shifts, often depending on whether the function of a part depends on its length, its area, its mass, or some product of these such as moment of inertia (mass times a length). The commonest comparisons are made between a particular dimension of a part on the one hand and a measure of overall body size such as mass on the other, and includes many familiar anatomical examples such as the relationship between brain volume and body mass, where the former scales at only about 0.7 times the rate of the latter. This is a case of negative allometry and the brains of larger animals are relatively smaller than those of smaller-sized relatives. Positive allometry is exemplified by the relationship between flight muscle mass and body mass in birds; the muscles must increase at a rate about 1.3 times the rate of increase in the body to maintain the lift force necessary in a flying animal. The interest of allometric evolution for evolutionary studies is that phenotypic integration is maintained in successive descendant morphologies despite the shift in the relative sizes and therefore also the spatial relationships of the parts. Like heterochrony and heterotopy, allometry should be treated initially as a description of a pattern of change in characters, and the question of how the integration amongst these characters is maintained, whether by developmental mechanisms or by selection, treated separately. The great majority of cases of allometric change affect low taxonomic levels, without leading to a radical morphological reorganization (Hanken and Wake 1993). However, there are a number of more extreme cases where the modifications are of sufficient magnitude potentially to have played a role in the origin of a new higher taxon.

The most frequently invoked possibility in this context concerns miniaturization and whether such extreme reduction in overall body size could lead to a radical new organization. There are many cases of modern miniaturized species associated with significant complex morphological changes involving both loss of characters and novel features (Hanken and Wake 1993; McNamara and McKinney 2005). Amongst the vertebrates (Rieppel 1996) are several modern salamander species, where miniaturization is associated with loss or incomplete ossification of several cranial bones, and lizards in which the disproportionate size of the otic capsule modifies the skull architecture quite extensively. In all these cases, however, the morphological changes are insufficient to accept the miniaturized forms as new higher taxa; they remain recognizable as merely extreme morphologies within the disparity of their respective groups.

However, there are several cases amongst the vertebrates where the fossil record does suggest an origin of a new higher taxon from a relatively, if not extremely miniaturized ancestor. The best documented are those of the birds and the mammals, both of which undoubtedly involved considerable body size reduction. In the case of the birds, the positive allometry between flight muscle mass and body mass meant that the evolution of muscles capable of generating the very high forces necessary for active flight could only have occurred realistically in a relatively very small bipedal dinosaur. In birds and mammals, the reorganization of the cranial anatomy consequent upon the relatively enlarged brain in animals of greatly reduced body size may have been important in allowing the absolute increase in brain size that occurred in these two taxa.

Amongst modern invertebrates there are some higher taxa which are serious candidates for an origin involving miniaturization. Rundell and Leander (2010) reviewed the interstitial marine meiofauna, consisting of miniaturized species of several invertebrate phyla with body lengths reduced to between about 60 μm and 2 mm. Even though still identifiable as members of familiar phyla, they often show quite extreme morphological changes, such as loss of the coelom in annelids, and reduction of the body musculature in favour of a ciliated surface for locomotion in several groups. The possible origin of the Phylum Loricifera by neoteny has been mentioned (Fig. 5.5), and the process must also have involved miniaturization to produce such small animals. How much of the morphological reorganization associated with the origin of this group was a direct consequence of the excessive size reduction is not clear, but it does seem as if the cells individually evolved smaller size in order to allow the complexity of such minute organisms (Sørensen et al. 2008). Another higher taxon generally accepted as a result of miniaturization are the mites, Acarinae, based on their general simplification of the arachnid body plan coupled with their small size.

In principle, the opposite evolutionary effect of gigantism could also have allometric effects leading to significant structural reorganization of the morphology. The most intriguing possibility concerns the rapid appearance in the fossil record of almost all the animal phyla during a relatively narrow window of 10–15 million years in the Cambrian, known as the Cambrian explosion, and described in detail in Chapter 7, section 7.2. The two possible explanations are either that there was indeed an astonishing and egregiously rapid origin of whole new body plans, or that the lineages all had a long, but recorded history. The latter view is supported by the molecular evidence, which consistently indicates divergences amongst the living phyla well before the start of the Cambrian period 540 million years ago. Despite this evidence, plausible Precambrian stem members of these modern phyla are virtually non-existent. One popular explanation for the inferred missing fossils is that the animals were all very small, soft-bodied meiofaunal or planktonic organisms lacking fossilizable skeletons. From this view, the Cambrian explosion consisted of a rapid evolution of large-bodied, relatively speaking gigantic forms of existing taxa that for functional reasons now required mineralization of the skeleton. Thus hard shells and cuticular exoskeletons evolved. Various suggestions have been made for the identity of the presumed selective pressure driving the increase in body size in parallel in many groups. Increased metabolic rates permitted by elevated oxygen levels is one, and an arms race between active predators and their prey another, not necessarily mutually exclusive. The increased body size would have resulted from a prolongation of the rate and/or the length of embryonic growth. This in turn would demand a number of allometric changes. As body size rises, the fall in surface area to volume ratio of the body creates an increasing need for specialized respiratory and excretory organs. There is also a tendency to reduce the strength to weight ratio of supporting tissues, making the evolution of a stronger skeleton necessary, while a shift from ciliary-based to muscle-based locomotion has to occur. Assuming that the metabolic

rate increased with the rising oxygen levels, then a higher rate of nutrient requirements would demand increasing specialization of the feeding structures. From the point of view of this scenario, such substantial modification of the morphology can be envisaged to the extent of resulting in a series of new higher taxa.

A less extreme, but equally feasible case of the origin of a higher taxon involving gigantism is the whales. Here, large body size achieved by gigantism is associated with major modification of the locomotory and physiological systems contributing to the extreme adaptation to secondary marine life. In this case, there is a fossil record illustrating this process, as described in Chapter 8.

5.3 Developmental mechanisms for the maintenance of phenotypic integration

The modes of morphological evolution just discussed, recapitulation, heterochrony, heterotopy, and evolutionary allometry, all consist of functionally coordinated patterns of evolutionary change in multiple parts of an organism, and give an indication of the importance of such complex, multi-trait evolution. Unfortunately, the developmental mechanisms maintaining this integration in the course of the changes in the several separate characters are still quite poorly understood. Certainly it would be mistaken to assume that a heterochronic or heterotopic modification only requires a simple change in an underlying developmental mechanism. As mentioned already, heterochronic change in the morphology can arise from a change in the starting conditions such as the size of the population of initial cells, rather than from a heterochronic change in a developmental process. Similarly, heterotopic morphology could result from events other than changes in the position at which a part originates in the embryo, such as differential growth of tissues around it causing it to shift position. For example, this occurs very clearly in the movement of one eye to the opposite side of the head during metamorphosis of pleuronectiform flatfish. Furthermore, there are other developmental causes of complex novel, but still integrated characters involving other processes than simply modifying the timing and spatial disposition of developmental events. Several complex vertebrate structures like the branchial arches and sensory nerves are associated with the appearance of a new embryonic tissue, the neural crest (e.g. Shimeld and Holland 2000), or of new kinds of tissue interactions such as induction of a light-sensitive organ by ectoderm as found in the eyes not only of vertebrates, but of many invertebrate taxa too (Shubin et al. 2009). What all these modes of multi-trait evolution share is the maintenance of functional and structural integration amongst the several parts during the course of evolution.

The mechanisms by which this is achieved are at the very core of major evolutionary change and are particularly relevant to the origin of new higher taxa which manifest large amounts of correlated changes in many parts of the phenotype. In so far as the phenotype is the product of its ontogeny, the proximate cause of all evolutionary change is modification of development, for which two ultimate causes exist. One is direct natural selection on the organisms, whereby those individuals which by chance develop better integrated traits survive: this aspect is discussed in Chapter 6. The other is pre-existing genetic regulatory mechanisms, in which integration is maintained by interactions amongst the developing cells, tissues, and structures themselves, via various modes of internal feedback. Mechanisms in this category may operate at one of several levels of ontogeny, classified perhaps as cellular- and tissue-level inductions, and modular processes at the molecular level.

5.3.1 Embryonic cellular and tissue interactions

The principles by which epigenetic interactions between developing tissues are responsible for generating functional integration amongst the parts of a structures have been amply illustrated by many examples, one of the most extensively studied of which is the tetrapod limb. For a given limb of a tetrapod to be functionally viable requires that the lengths of the bones, the positions of the joints between the bones, the anatomy of the muscles that attach to the bones, the arrangement of the motor and sensory nerves serving the muscles along with their connections within the spinal cord, and the limb vascularization must all match one another. The necessity for this outcome is obvious, and it is achieved by a number of self-regulatory or feedback mechanisms between the different cells types and tissues within the developing limb bud (Wolpert et al. 2011). The precursors of the bones occur as condensations of cells that form cartilage, differentiating from proximal to distal as the limb bud grows out. The joints between the individual elements develop as cartilage-free zones as a reaction to stresses created by developing muscles. Meanwhile, muscle cells enter into the limb bud by migration from the somites. Their patterning is determined by the connective tissue framework associated with the skeletal elements to which they become attached, each individual muscle differentiating correctly in direct response to signalling from its own particular associated skeletal element. In contrast, motor neurons are already specified as to which muscle they will innervate while still within the nerve cord. They grow peripherally under the guidance of several kinds of molecular signals given out by the target muscle cells, some attracting and some repelling the growing tip of the axon. By this means, an axon extends distally and attaches to the correct muscle cell. A larger number than are required actually make contact and form neuromuscular connections, but then the majority die, leaving just a single neuron per muscle fibre. Finally, appropriate vascularization occurs as blood vessels condense within the pre-existing connective tissue. The picture that emerges from this somewhat oversimplified account is one in which molecular signalling between the different parts of the developing limb ensures that integration amongst the participating cells and tissues is achieved in the final product. Given the mechanisms behind this pattern of interaction, it is predictable that a mutation affecting the structure will usually affect all the parts appropriately. An enlarged limb bone, caused perhaps by an increase in the period of expression of a regulatory gene during the initial expansion of the limb bud, will directly induce the development of appropriately modified muscles, nerves, and blood vessels. Thus the limb as a whole will remain a functionally viable structure.

Tissue-level feedback processes like this are universal amongst organisms, and offer a ready account for integrated changes in a defined structure. In principle, they could account for modifications that are significant enough to count as evolutionary innovations, such as limb reduction in certain skink species and limb elongation in giraffes. However, within a lineage that culminates in a more comprehensive morphological transition as is characteristic of a new higher taxon, there are evolutionary changes in non-contiguous parts of the organism that are also functionally integrated but that cannot feasibly result from direct interactions between cells and tissues during ontogeny. Fore-limb, hind-limb, and cerebellum organization in terrestrial tetrapods; metabolic rate, heart structure, and dermal insulation in endotherms; form of feeding mouthparts, detection of flowers, and social communication in bees: a thousand obvious examples spring to mind. Clearly tissue-level epigenetic interactions cannot be directly responsible for the maintenance of phenotypic integration amongst such disparate and distantly spaced parts and therefore the extent to which these processes contribute to major evolutionary change is limited. There must be additional mechanisms.

5.3.2 The role of molecular genetic mechanisms

The cell and tissue reactions involved in the development of phenotypic integration and therefore ultimately its maintenance over evolutionary time are mediated by the various kinds of molecular signalling mechanisms between cells that occur during development. Fully understanding these processes requires opening the 'black box' between the gene sequences and the adult phenotype, the epigenetic machinery, and this is work still very much in progress. Several organizational principles underlying the genetic regulation of development have been discovered which in a number of respects proved unexpected, and which have profound implications for explaining major as well as minor evolutionary transformation.

Amongst the first such principles to emerge as a result of molecular analysis was the phenomenon of cis-regulation, whereby proteins, referred to as transcription factors, are produced by several different regulator genes all of which affect the same target gene, via an adjacent part of its DNA, referred to as the transcription factor binding site (Fig. 5.7a). Some transcription factors are activators, some are repressors, and their activities combine to determine the place, the timing, and the level of transcription activity of the target gene. This latter may be a gene coding for a functional protein molecule, or may itself be a regulator gene producing its own transcription factor for use elsewhere. A second organizing principle of the genetic control of development is that a group of homologous interacting genes and gene products is frequently found to act

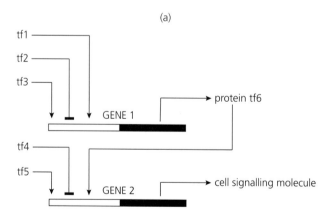

(a)

Figure 5.7 (a) The principle of cis-regulation of gene activity. The action of a regulator gene may be activation or inhibition and the activation state of a gene depends on the integration of all the regulator genes. (b) The Hedgehog cell-signalling module, an example of a genetic developmental module conserved throughout the animal kingdom and used in numerous developmental events throughout the organism. Above: in the absence of the HH-N element, the membrane protein PTC inhibits SMO and the complex of molecules inhibits transcription of the HH target gene. Below, when HH-N attaches to the membrane, SMO is longer inhibited and it promotes activity in the complex that leads to activation of the target gene. (From Ingham et al. 2011, reproduced with permission from Nature Publishing Group.)

(b)

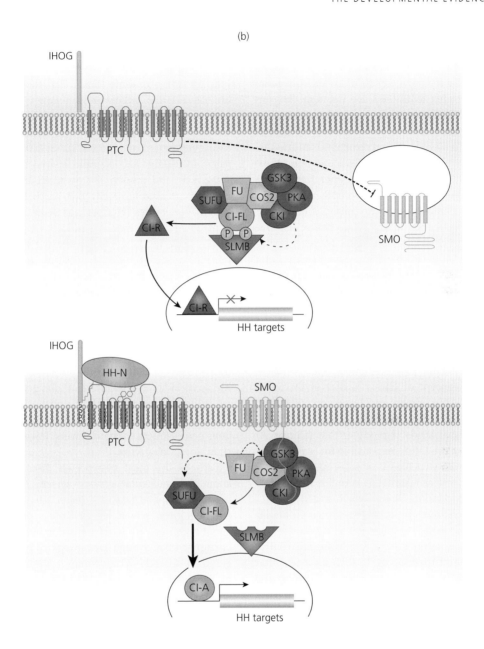

Figure 5.7 *(continued)*

together at several different places and times within a developing embryo, and thus to constitute a molecular module affecting the development of what may be a number of entirely different, functionally remotely related parts of the embryo (Fig. 5.7b). A third principle is the hierarchical organization of the genetic developmental system responsible for the development of a particular structure, with higher-level, or upstream elements affecting lower-level, downstream elements, and these in turn controlling the lowest level of the hierarchy which consists of genes producing the protein molecules immediately responsible for the actions of cells. The hierarchy is certainly not strict, and cases of reciprocal effects between genes at the same hierarchical level, and recursive effects of lower-level genes acting on higher-level ones are frequent. Salazar-Ciudad (2009), for example, has criticized an oversimple hierarchy model, pointing out the extent to which a layered hierarchical structure is blurred by elements that act at different stages and in multiple locations in the same embryo, depending on the molecular environment and other epigenetic factors such as mechanical stresses on cells.

These discoveries about the nature of the molecular-level regulation of development are encapsulated in the idea of the Genetic Regulator Network (GRN), a very helpful concept for unravelling the molecular basis of development: a GRN is a modular, multilayered, hierarchically organized system of genes and gene products. The elements of a GRN interact to regulate the genes that produce the proteins ultimately responsible for the activities and metabolic processes of the cells. The most structured concept of GRNs is that of Davidson and Erwin (2006; Erwin and Davidson 2009) who have described the architecture in terms of a number of different kinds of modules that interact. At the centre of the GRN responsible for a particular morphological structure is a sub-circuit or module they refer to as a kernel. A kernel is responsible for the transcription factors which specify the region, or domain of the embryo where a particular structure develops. Notably the individual genes constituting a kernel have a highly specified interaction amongst themselves, to such an extent that malfunctioning of any one of them is likely to disrupt the functional failure of the whole kernel and therefore of the initiation of development of the part of the organism for which it is responsible. For this reason, kernels are very highly conserved amongst distantly related higher taxa. For example, specification of the heart domain in *Drosophila* and in vertebrates is by a very similar regulatory gene sub-network, notwithstanding the completely different kinds of hearts that develop. The second category of modules they refer to as 'plug-ins' by analogy with computer programmes. Plug-ins are also conserved amongst even distantly related taxa and function as signalling systems in a wide variety

of different regulatory contexts, rather than being responsible for any particular part of the embryo. One of the best known such modules is the Hedgehog signalling module (Fig. 5.7b), parts of which are represented in choanoflagellates, poriferans, and cnidarians (Matus et al. 2008). The full module occurs in all animal phyla, where it is involved in the development of a large array of different structures including for example segmentation, gut, wing, eye, and muscle in *Drosophila*, and somites, gut, smooth muscle, spinal cord, limbs, brain, bone, etc., in vertebrates (Hooper and Scott 2005; Matus et al. 2008). Third, there are modules that act as input/output (I/O) switches, which act to repress or permit the activity of other sub-circuits in particular parts of the embryo. Finally, at the lowest level of the GRN come the differentiation gene batteries. These are genes whose protein molecules are responsible for specific cell activities such as cellular differentiation and movement. They are subject to the upstream cis-regulation by the transcription factors of the I/O switches and plug-ins at the higher hierarchical level. The differentiation gene batteries are evidently the most evolvable parts of the overall GRNs, because differences are found in the homologous gene sequences even between lower-level taxa right down to species.

Perhaps the most unexpected discovery of all about the molecular genetic system for regulating development was this degree of conservation amongst taxa, which colours any considerations about major evolutionary transition and the origin of higher taxa. A high percentage of homologous genes are shared not only amongst the eumetazoan phyla, but also to an extent with the more basal groups, cnidarians and poriferans. Indeed, all of the main four signalling pathways of eumetazoans, Wnt, Notch, TGF-b, and Hedgehog are represented in sponges, as are Hox genes. In the non-bilateral organisms, these regulatory genes are obviously not associated with the development of such traits as bilateral symmetry, segmentation, or organs such as hearts and brains with which they are associated in the Bilateria. Rather, they have functions associated with ancestral character states of tissue organization and physiological functions of cells. The implication is that a great deal of the molecular developmental machinery of bilaterian animals arose by new functional interactions amongst the pre-existing regulatory genes of non-bilaterian metazoans, leading to their reorganization as increasingly discrete modules. Furthermore, a significant percentage of metazoan regulatory genes and transcription factors are also found in micro-organisms such as the choanoflagellates, the probable sister-group of the Metazoa. The underlying essential genome of all organisms is indeed ancient.

The different degrees of conservation of the different levels of the regulatory hierarchy lead to speculations about the nature of the evolution of higher taxa. When differences in the patterns of segmentation and appendages in arthropod taxa were first shown to correlate with the patterns of expression of the regulatory gene family of Hox genes, the thought was sometimes expressed that perhaps simple genetic changes, even in the extreme a single mutation in a regulatory gene, might cause a large but coordinated evolutionary change, creating a 'hopeful monster'. The idea that such a process could be responsible for the rapid generation of the body plans of the different phyla that appeared in the Cambrian was superficially attractive, but such a naive view was rapidly extinguished as the complexity and numbers of elements actually involved in determining the basic organization of a eumetazoan became apparent. Nevertheless, the different degrees of conservation of modules within the hierarchy of a GRN do suggest that certain parts of the genome have different evolutionary properties to others. The very high degree of conservation of kernels leads to the implication that not only are they of great phylogenetic age, but also that their origin was associated with the very origin of the different body plans of the animal phyla (Davidson and Erwin 2006; Erwin and Davidson 2009; Peter and Davidson 2011). Once the kernels had acquired the very strong functional integration amongst the genes and gene products of which they are composed, further evolutionary change in the kernels without loss of that integration and hence loss of function is deemed to be extremely improbable. Divergence of the stem lineages from the ancestral bilaterian and therefore the origin of the different phyla, and perhaps classes too, occurred primarily by mutations of the genes of the somewhat less conservative plug-in and switching modules, where

mutations affecting the timing and locality of their deployment within the embryo changed. These modifications in turn affected the differentiation gene batteries at the lower levels of the hierarchy which are responsible for the appearance during development of the phenotypic structures. Finally, the lowest taxonomic level differences, typical of those seen between sister-species, arose from finely tuned modifications to the patterns of cell differentiation, organization, and physiological properties due to mutations in the most downstream differentiation genes, which are seen to be the least conserved.

This simple model of evolution of genetic regulatory systems that assumes different levels of evolvability at different points of the regulatory networks is consistent with the different degrees of conservatism actually observed amongst the different kinds of elements of the GRNs of modern organisms. It also offers insight into the problem of how to reconcile the minor genetic differences between lower-level taxa such as sister-species with the major genetic differences between higher taxa, when the higher taxa themselves are presumed to have evolved by a gradual succession of speciation events. Mutation of a structural or housekeeping gene or one of its immediate cis-regulatory genes may cause minor phenotypic variation, such as in the size or the physiological properties of a part, variation only to the extent characteristic of the heritable differences between species and which is susceptible to fixation by normal natural selection, or genetic drift. Occurring at the lowest level of the GRN hierarchy, this new genetic effect is likely to be manifested only at a late stage in development. The question is how a succession of unit evolutionary steps of this nature could accumulate in such a way that results in the large differences characterizing higher taxa, that are due to genetic effects manifested at earlier stages in development, and that have a cascading effect on the subsequent stages of development. It is not immediately obvious how small-effect, late-acting mutations could shift the timing of their expression to an earlier stage in development, or how they could combine with one another in some sense to cause large changes in the adult phenotype. The GRN framework avoids this requirement by proposing that it is mutations of genes in modules already acting at the higher levels of the GRN hierarchy that cause the large evolutionary changes, because they already act early on in development and so potentially have larger effects initially. A beneficial mutation of a gene that is part of a kernel, rare as that might be, may be imagined to cause a shift in the characteristics of a major part of the phenotype such as the segmentation pattern of an arthropod, or a mutation in a cell-signalling plug-in module to affect the timing of development and hence size of several parts of an embryo simultaneously.

However, this explanation has distinct echoes of the 'single-mutation, hopeful monster'. It raises the paradox of evolvability, namely how large evolutionary change in a complex system of functionally related parts is possible without loss of overall integration of the system as a whole. In this case, the system in question is the genetic developmental module, whose correct functioning depends on the exact interactions amongst the several integrated genes of which it is composed. Logically this is the same problem as for phenotypic evolution, discussed at the start of Chapter 3: the complexity of an integrated system promotes constraint on change while evolvability requires change. Three conceptual solutions were offered for resolving the evolvability paradox of phenotypes. One of them is developmental feedback of some kind, in which a change in the development of one structure or tissue automatically adjusts the development of others so that together they maintain appropriate functional relationships. In the case of the highly conserved parts of genetic developmental modules, a comparable feedback mechanism would require a mutation in one of the genes to cause compensatory mutations in others, for which no possible mechanism is known to exist.

The second phenotypic solution in principal is modularity, to the extent that a semi-independent module within a system consists of many less parts than the whole system, and therefore the chance of a change in one of the parts of the module still fitting in harmoniously with the rest is higher. In the case of the developmental regulatory system, the modular structure may well be a major factor that increases evolvability in the less-conserved, lower-level modules in the hierarchy. However, the higher level of conservation of the kernels and plug-ins,

suggests that in modern organisms the integration of these modules has proved too strong for more than minor evolutionary changes within them to be allowed without loss of fitness.

The third theoretical explanation of evolvability at the phenotypic level is the process of correlated progression of traits (Chapter 3, section 3.3), in which any one element of the complex system can undergo a small change without significantly reducing its integration with the rest, but further evolution of that element requires appropriate small, correlated changes in other elements to accumulate sequentially. A similar concept offers an explanation for how a mutation in the genetic regulatory system may have a small phenotypic effect yet act early in development, and how such mutations may accumulate until collectively they cause phenotypic modifications of the magnitude that distinguishes separate higher taxa. Consider a regulatory module high in the hierarchy of the GRN which consists of several genes whose products are transcription factors both of one another and of the gene that produces the functional output molecule (Fig. 5.8). The system is highly integrated and a mutation with a large effect in any one of the genes is likely to cause the system as a whole to fail. However, a mutation in a gene that has a small effect, one that slightly modifies its output, may have but a small effect on the cis-regulation of the other genes and therefore on their outputs. This will include the functional output molecule, whose timing, extent, or place of expression in the phenotype will be slightly modified. The result is a small modification to the adult, perhaps to the extent of a species-level difference, but caused by a high-level, early-acting genetic module. A second mutation may similarly occur and so on. The end result could easily be a large shift in the action of the module, and a modification in the phenotype characteristic of a new higher taxon. Yet throughout this process of gradual evolutionary change, the regulatory genes involved act at an early stage of development. This is in contrast to a supposed small, species-level modification caused by a mutation in a gene in the lower part of the hierarchy. Of course such mutations occur, can occur sequentially, and are responsible for many, perhaps most low taxonomic-level differences. However, what they probably cannot do is become

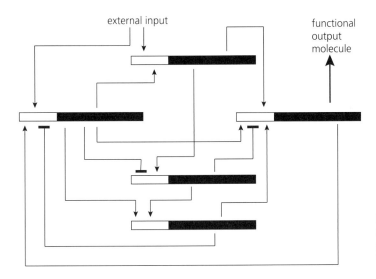

Figure 5.8 Model of a highly conserved genetic module. See text for explanation. The bars are genes, with the transcription factor binding sites white. Arrows are excitation, bars are inhibitions.

higher-level, earlier acting elements, and therefore part of the regulatory system that is responsible for laying down basic body plans of higher taxa.

In the light of these ideas about the evolution of the genetic regulatory system, it is also possible to address the question of how highly conserved modules arose in the first place, and thus to create a feasible scenario for the origin of the animal phyla associated with the Cambrian explosion. As several authors (He and Deem 2010; Marshall and Valentine 2010; Lorenz et al. 2011; Erwin and Valentine 2013) have stressed, explanation of the origin of the molecular regulation underlying the body plans of animal phyla cannot be divorced from either the pre-existing mechanisms already present in stem Metazoa and Bilateria as inferred by comparisons with modern representatives, or from the ecological opportunities for diversification during the immediate Precambrian and Early Cambrian indicated in the palaeoenvironmental signals discussed elsewhere (Chapter 6, section 6.5). First, it may be imagined that relatively loosely integrated regulatory genes existed in the pre-metazoan multicellular grade, whose activation was to a large extent environmentally induced and whose role was to control the transcription of genes determining cell activities such as differentiation, movement, and adhesion. The resulting phenotype consisted of a variety of cell types arranged as layers and tissues, but the precise arrangement of these was not very critical, as the

organism's vegetative functions were by diffusion and ciliary action rather than by discrete organs. New environmental conditions such as increased oxygen and calcium availability created novel ecological opportunities for more specialized ecotypes such as bulk suspension feeders, active browsers, and faunivores, and thus the potential arose for a diverse community of several guilds of more active, specialized organisms. Realization of the potential required more precise adaptations on the part of each taxon in the face of developing competitive and nutritive interrelationships of diverging members of the community, a challenge taken up, so to speak, by the genetic developmental network. Occasional random gene duplications followed by mutation in paralogous genes is a common way in which new genes arise. Such paralogues of existing regulator genes are likely to produce similar but not identical transcription factors acting on the same target genes, and causing small but significant changes in their expression. For example, a mutated version of a regulator gene copy could be imagined as having a slightly enhanced effect in one of its target cells, thereby advancing the time of transcription and so causing an increase in the number of the cells for whose production it is responsible. If the molecular effect is small, it need not disrupt the overall functioning of the cis-regulation of the target gene but only modify it slightly, and the consequential phenotypic variant may be positively selected. By this process, more and more changes in the pattern of transcription arise without loss of integration of the regulatory system. Meanwhile, the greater and greater is the phenotypic effect. Thus, a sequence of mutations each having small individual effects may be selected. However, as they occur in genes that already act at an early stage in development, collectively they result in the large phenotypic differences characteristic of higher taxa. Furthermore, one attribute likely to be selected is increasingly tight integration amongst the individual regulatory genes. This is because as the successive phenotypes shift from loosely arranged tissues to more complex, organ-based structure in order to take advantage of new roles within the increasingly complex community, so more precise timing and regionalization of specialized cell types is required. The outcome is the relatively highly conserved groups of interacting genetic elements, the modules, each responsible for the correct assembly of the differentiated parts of the organism.

Three decades ago, when very little was known about the principles of the molecular regulation of development, Gould (1989) noted that the very rapid appearance of new animal phyla in the Cambrian was followed over the next few hundred million years by extensive evolutionary diversification of these body plans but that no new ones originated. He believed that this phylogenetic pattern reflects a property of the evolving genomes, and assumed that the new evolutionary radiation consisted initially of a sort of 'experimental' phase during which the genome could vary readily and a wide range of different body plans rapidly evolved. This phase was then followed by the genome becoming increasingly more internally structured and less susceptible to major adaptive change, preventing new body plans from appearing. He referred to the phenomenon as 'congealing'. In Gould's eyes this was a general mechanism, applicable not only to the Cambrian explosion of animal phyla but in principle to the origin of all major new groups with distinctive body plans, which would include the classes of invertebrate phyla and the highest vertebrate taxa. The concept was widely criticized on the grounds that there was no feasible mechanism known for congealing, and in any case the hypothesis implied a heterogeneous nature of genomic evolution when

most evolutionary biologists accepted the adequacy of simple Darwinian natural selection of mutations of small effect to account for all phylogeny at every taxonomic level. However, the emergence of the same general idea as 'congealing', but now on the basis of the empirical evidence concerning the molecular nature of genetic regulation, and in particular the different degrees of conservation amongst different kinds of modules, resurrects Gould's hypothesis. In this light, the early stage in the origin and radiation of new higher taxa consists of the evolution of the higher level modules of the genetic regulatory system from a relatively more flexible state via a correlated progression mechanism of small genetic mutations. Once established as more highly organized, the level of integration within these modules had increased to a point at which further changes were unlikely to be viable. Meanwhile, the less tightly integrated lower-level elements of the networks remained more evolvable thanks to their late action and smaller overall effect on development. These were therefore responsible for the diversification of the phyla and classes into their numerous lower taxa.

Current empirical evidence bearing on the molecular basis of actual major evolutionary transitions is extremely limited, and consists mostly of comparisons between the patterns and timings of the expression of regulator genes in the developing tissues of homologous parts of related higher taxa. The methodology is to infer from this a hypothetical common ancestral regulatory system, and then to suggest what changes in regulation might have occurred in the respective diverging lineages. One of the most studied examples concerns the distribution of Hox gene paralogues amongst the major taxa (e.g. Pick and Heffer 2012). This can be compared to the differences in antero-posterior differentiation and segmentation of the body axis in the different phyla (Fig. 5.9).

Most studies of the genetic regulatory differences between higher taxa do not concern overall body plans but are necessarily limited to single structures or parts. For example, in the case of the origin of the turtle carapace described earlier (see section 5.2.3 this chapter), the regulatory genes expressed by the carapacial ridge, which is a novel structure unique to the chelonians, include initially

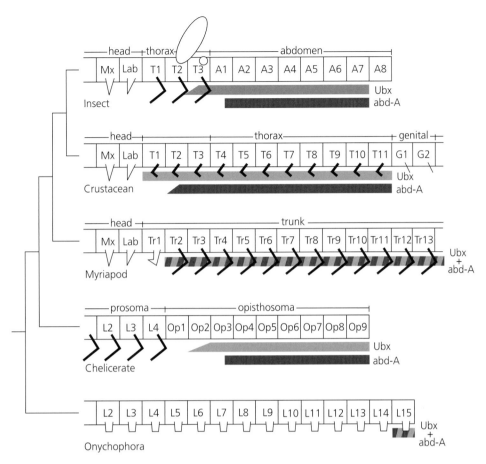

Figure 5.9 The region of expression of two Hox genes, Ubx and abdA, in different arthropod classes. Note that the anteriormost expression of Ubx coincides with a significant morphological change in segmental morphology. (After Carroll et al. 2001 reproduced with permission from John Wiley & Sons.)

those associated with the formation of the dermal part of the embryo. Subsequent development of the carapace involves the expression of regulatory genes that are involved in the histologically somewhat similar apical endodermal ridge of tetrapod limb buds (Moustakas 2008). This suggests that the evolution of the carapace involved the recruitment of a module of the limb-forming genetic regulatory network to a new region of deployment around the body margin. The consequence was the heterotopic shift in development of the dorsal ribs, and the appearance of the novel ossification pattern associated with them (Nagashima et al. 2009).

Another widely studied case concerns the genetic basis of the transition from a fish fin to a tetrapod limb, where differences in expression patterns of homologous and paralogous Hox genes in the development of the appendages of teleost fishes and living tetrapods (Fig. 5.10) suggest possibilities for what may have been the genetic basis of the evolutionary change (Davis 2013; Yano and Tamura 2013).

Important research steps as such studies are, they are as yet too broad, and too limited to comparisons of when and where the components are active in different modern taxa, to throw much empirical light on hypothetical intermediate stages. Eventually enough will no doubt be known about the principles of GRN function to be able to infer these. At the present time, what is known suggests first that the modular architecture of gene regulation is

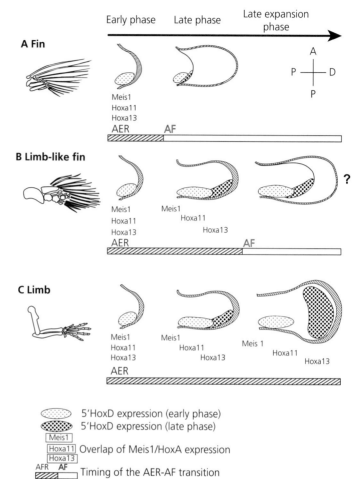

Figure 5.10 Expression patterns of some of the genes involved in appendage development in vertebrates. In fin development, the apical endodermal ridge (AER) which controls proximal-distal differentiation is short-lived and replaced by an apical fold (AF) in which the fin rays develop. The expression of the Meis, Hoxa11, and Hoxa13 Hox genes ceases. In the development of the tetrapod limb, the AER is not replaced by an apical fold but remains active and 5'HoxD enters a late phase of expression. The genes Meis1, Hoxa11, and Hoxa13 that are active in the AER are activated in succession and are associated with the development of the stylopod (humerus/femur), zygopod (radius–ulna/tibia–fibula), and autopod (wrist–hand/ankle–foot) respectively. Also the expression of shh in the distal limb expands and is associated with the width of the autopod and the number of digits. (After Yano and Tamura 2013 reproduced with permission from John Wiley & Sons.)

indeed important in facilitating significant modification to phenotypic structure while maintaining integration of the resultant phenotype. Second, the mechanism of cis-regulation, whereby the time and place of expression of the genes directly dictating cell behaviour are controlled by numerous upstream regulatory genes each with small effect, points to a mechanism for accumulating a long enough sequence of small-acting mutations to lead eventually to a large phenotypic consequence.

5.3.3 Phenotypic plasticity

Innumerable studies on all kinds of organisms and traits have demonstrated how common it is for the

development of an organism to be modified in response to an environmental cue in such a way as to result in a phenotype better adapted to that environment. There are also experimental studies demonstrating that artificial selection of those individuals that most readily or extensively show the phenotypic response can lead to the situation where the environmental cue is no longer necessary for the adaptive response to occur. Instead, the selection has produced individuals in which the adaptive response has become entirely genetically determined. This is the process known as genetic assimilation (Waddington 1953), and there has been a great deal of speculation and argument about how significant it is as a natural evolutionary process (reviewed for example by West-Eberhard 2003, Pigliucci 2010). Some have downplayed its significance, in the belief that it conflicts with Darwinian natural selection because the variation upon which selection acts is not random but already adaptive. Indeed, genetic assimilation has been mistakenly conflated with Lamarkian inheritance of acquired characters on occasion, although of course the very capacity to respond adaptively to direct environmental cues is itself presumably a result of natural selection. Others in contrast regard phenotypic plasticity as a perfectly viable mechanism for speeding up adaptive evolution by natural selection, and quote in its favour cases of genetically determined features in certain species that in related species are environmentally determined. Phenotypic plasticity can, as it were, lead the way by generating instant adaptive change within a population and therefore enhancing its survival in a new environment, to be followed in due course by natural selection in the population favouring random genetic mutations that tend to preserve and enhance the new morphology. There is particular interest in the idea of the role of behavioural plasticity, or learning as an initial step preceding morphological evolution. Neufeld and Palmer (2011) for example, considered the not infrequent situation where individuals of a species exposed to a new predator develop a predator-resistant morphology such as a thicker shell in some mollusc species, or a larger tail in certain anuran tadpoles. The first stage is necessarily the behavioural one of detecting and interpreting the risk posed by the new predator to which they are exposed, followed by the developmental response of morphological modification. Those individuals which respond behaviourally most quickly receive the protection of the morphological response earliest, and therefore are selected. Modelling this situation demonstrates, unsurprisingly, that the greater the facility to learn that the predator is present, the faster the morphological response, and presumably therefore the higher the rate of genetic assimilation in a population continuously exposed to the same predator.

There seems little doubt that phenotypic plasticity plays a role in many cases of evolutionary change at a low taxonomic level, certainly within populations and sometimes perhaps as an initial stage in speciation (reviewed extensively by West-Eberhard 2003). However, the present question is whether it is ever more than a microevolutionary process producing relatively minor adaptation, or whether it could be significant in major evolutionary transitions and the origin of higher taxa. West-Eberhard (2003) suggested that the origin of tetrapods included an element of phenotypic plasticity because the lineage went through a phase of facultative air-breathing phenotypes which still possessed gills and aquatic respiration. She speculated that such an intermediate species in the lineage would be capable of a plastic response to the oxygen concentration of the water. Those exposed to a lower concentration developed larger lungs and smaller gills and therefore a higher air-breathing capacity. They would have been better adapted to survive in parts of the habitat with poorer oxygenated water. In a population of the species living permanently in this region, genetic assimilation of the development of the enhanced air-breathing capacity followed, and so a step further along the gradient towards terrestrial life taken. While such a scenario may be feasible, it is impossible in practice to distinguish it from the more conventional natural selection scenario in which those individual variants of the species which happened by chance to be better air-breathers could survive in the lower oxygenated part of the range. Furthermore, phenotypic plasticity cannot account for the range of other biological modifications which are also necessary for a significant shift towards the terrestrial habitat. In this, and probably in all comparable cases of higher taxonomic level transitions, phenotypic plasticity is a feasible microevolutionary process for producing relatively small changes in limited parts of the organism, but one that has no other particular role in the origin of higher taxa.

There is one possible special case where it has been proposed that phenotypic plasticity played a uniquely important role. Newman and colleagues (Newman et al. 2006; Newman 2010) have argued that the Metazoa itself, along with the the basic body plans of animals, originally arose as a consequence of far more direct responses by an ancestral cell mass to mechanical and chemical environmental cues, rather than as a result of selection of random genetically determined variants. They noted that several processes involved in development, such as cell adhesion, molecular diffusion, and molecular-level activity oscillations can be affected by environmental conditions, even though today they are under genetic control. The authors speculate that such environmental determination was much more important during the origin and early diversification of metazoans. Differing environmental conditions acting on ancestral cell masses could result in a variety of different morphological patterns of cell differentiation, epithelia, cell layering, internal compartmentalization, serial repetition of parts, etc. Subsequent natural selection of modified existing regulator genes then led to modern genetic networks evolving from the primitive unicellular genetic system, and gradually taking over control of the development, stabilizing those particular morphologies best adapted for the early environment. Thus what had started off as a series of interchangeable morphotypes, each determined by the particular immediate environmental conditions, evolved into the variety of different body plans now primarily controlled genetically. However, even if this particular scenario of what amounts to early genetic assimilation is correct, the process applied only to the initial, Precambrian phase of metazoan divergence.

Another special case that may be regarded as phenotypic plasticity is where a species possesses a larval form, or sequence of larval forms, followed by metamorphosis to adult morphology. The same genome is responsible for the different stages, and the trigger for the developmental transition from one to another is usually environmental, such as ambient temperature, day length, or humidity. From this aspect, a major evolutionary transition involving neoteny, the retention of the larval form to sexually mature adulthood by permanent suppression of metamorphosis (see this chapter, section 5.2.2), would be a case involving phenotypic plasticity. Possible examples of the origin of a new higher taxon involving neoteny were discussed earlier and include vertebrates, Loricifera, and perhaps others.

Apart from these rather special cases, there is no evidence that phenotypic plasticity plus genetic assimilation plays a role other than as a process of microevolution.

5.4 Summary

The evidence from studies of development indicates the existence of significant intrinsic mechanisms for maintaining phenotypic integration during evolutionary transitions. Empirically, numerous well-studied examples where induction of one tissue by another, or feedback processes by which different associated tissues self-adjust to maintain appropriate relationships to one another are familiar. All are facilitated by molecular or mechanical signalling of many sorts. At the molecular genetic level, the gene regulatory system has its own mechanisms, notably upstream cis-regulation allowing moderation of the genetic effect of a mutation so it need not deleteriously affect the viability of the system, and modular architecture allowing already well-integrated developmental genetic modules to be deployed as a whole in new parts of the embryo, generating novel structures.

Phenotypic plasticity is another category of mechanism that in principle can maintain integration during evolutionary change, although only as a manifestation of the developmental processes just mentioned: the modification to the phenotype may be environmentally induced, but the resulting phenotype's overall integration still depends on the cellular and genetic processes that exist in the unmodified version.

These categories of mechanism are, however, limited and cannot be the sole, perhaps not even the prime maintainers of integration of the phenotype over the course of the major transitions leading to new higher taxa. Tissue interactions can only exist in contiguous parts of the phenotypic structure, such as discrete organs. They may play a part in keeping a limb functional during its evolution, but they cannot keep the rest of the organism appropriately tuned to the modified action of that limb. The genetic modules are also limited in effect because they can only affect discrete parts of the organism. They may assist in the coherent evolution of the arthropod segmentation, the turtle carapace, or the tetrapod limb, but again they cannot affect the relationship between these particular parts and the rest of the organism. To look for mechanisms that can maintain appropriate functional and structural relationships amongst all the parts of the phenotype over major evolutionary transitions requires consideration of control by natural selection as the lineage undertakes its long trek through morphospace, affecting numerous parts of the phenotype.

CHAPTER 6

The ecological perspective

Of all the aspects of the origin of new higher taxa and the long treks through morphospace that this entails, none is less open to direct empirical evidence and therefore more open to speculation than the nature of the ecological circumstances involved. There have been innumerable, often highly imaginative suggestions about the supposed selective force that 'drove' some particular major transition, such as Lull's (1918) classic invocation, developed by Romer (1958), of selection for the ability to disperse to more permanent bodies of water during the dry season to account for the emergence of the tetrapods onto land, or the widespread belief that the evolution of the exoskeleton in various invertebrate phyla was an adaptive response to a rise in atmospheric oxygen levels in the early Cambrian that permitted increased body size. However, discussions are scarce of any general principles about what causes an evolving lineage to follow a particular morphological direction affecting numerous characters over a sufficient distance that a quite new kind of organism, representing a new higher taxon, results. Does the lineage track something akin to an ecological gradient, in which case what kind of gradient could be so persistent, and so multidimensional that it induces consistent selection for many characters? Or is the driving force closer to a random Brownian or Markovian-like process in which the direction of each step is random with respect to the previous step? If so, the trend may be no more than the summation of the responses to a temporal sequence of different, transitory selective forces each acting on a single or a few characters at a time? A process of this nature might by chance occasionally produce what appears to be a largely deterministic, unidirectional long-term trend. There is also the

question of why exactly there are frequently short-term radiations superimposed at various stages of the overall lineage, as revealed by the fossil record and also by molecular divergence date analyses of modern taxa. Sections of the lineage may indeed be a very bushy structure in close-up view, and the lineage as a whole only be apparent as a more consistent long-term trend from a more distant perspective.

The interaction between the evolving lineage and its environment is also relevant to the maintenance of functional integration amongst the many changing parts, from ancestral to highly derived phenotypes. In the previous chapter, possible ways in which the integration may be maintained by developmental processes were discussed and it was evident that there are limits to the extent to which such mechanisms can operate. For more major transitions, affecting more parts of the phenotype and to a greater biological extent, additional processes must be at play, of which natural selection acting on multiple, integrated functions is the prime candidate.

Empirical ecological evidence relevant to these questions concerning the origin of new higher taxa is hard to come by. The field of evolutionary ecology, as understood by neontologists, considers the relationship between environment and evolution little, if any, further than the ecological circumstances associated with speciation. Whether there are any longer-term and higher-level processes involved in long sequences of successive speciation events is beyond its range. On the other hand, the main focus of palaeoecology has necessarily been on the observable patterns of ecological and community change on a geological time scale and at a supraspecific level, usually genus or family, rather than on trends in individual lineages. It has therefore tended to be

The Origin of Higher Taxa, First Edition. T. S. Kemp.
© T. S. Kemp 2016. Published 2016 by Oxford University Press and The University of Chicago Press.

concentrated on the causes of long-term taxonomic turnover. While this is on the right temporal scale for considering major evolutionary transitions, there are formidable epistemological barriers to knowing much about the detailed environmental conditions under which such evolutionary events occurred. First, there is the taxonomic incompleteness and low temporal resolution of the fossil record, limiting how much can be revealed about the community structure within which the transition took place. Second, despite considerable advances in geochemical analysis of past conditions, there is still a very limited amount of information available about abiotic features of palaeoenvironments and the selective forces they imposed. Third, like the ancestral stages themselves, the exact habitats and geographical regions that they occupied are almost invariably unknown, and can only be inferred from the broad conditions appertaining at the time. Thanks to this lack of direct evidence, reasoning about how an extensive evolving lineage interacted with its environment therefore consists mainly of inferences from what has been concluded about the pattern and functional significance of the evolution of the phenotypic characters.

In Chapter 2, the metaphor of a multidimensional adaptive landscape was employed in order to help clarify the meaning of 'higher taxon'. As described there, the two general categories of adaptive landscape are Sewell Wright's (1932) original version in which the coordinates represent gene frequencies, and Simpson's (1944) version in which the coordinates represent the values of phenotypic traits. The topography, peaks and valleys, represents the fitness values respectively of particular gene or phenotypic trait combinations. Adaptive landscapes have been criticized numbers of times for not being a sufficiently realistic representation of the biological world (a fate inevitably suffered to some degree by all metaphors, otherwise they would not be metaphors at all, but the real thing). An adaptive landscape should only be considered as a visual representation of what speculation, evidence, or hypothesis implies, rather than as a predictive hypothesis about the world in its own right. With this proviso, an adaptive landscape can be used to represent not only the relative fitness of different combinations of traits, but also various dynamic

aspects of evolving lineages. A lineage can be described as traversing a landscape in a manner that corresponds to the presumed relationship between the rates, magnitudes, directions, and limits of trait evolution, and the patterns of selection forces applied. It is thus possible to create the kind of adaptive landscape that is illustrative of the particular pattern of character evolution associated with the origin of a new higher taxon. From this point, whatever palaeobiological evidence there is that bears on actual cases can be considered, to see how far it is consistent with the ecological circumstances of major evolutionary transitions inherent in that particular form of adaptive landscape.

6.1 The correlated progression model and adaptive landscapes

The proposal presented in Chapter 3 that the correlated progression model of the phenotype and its evolution offers the most realistic account of major evolutionary transitions entails a number of implications about the nature of the selection regime imposed by the environment on the evolving lineage. The first is that most traits of a phenotype are subject to genetic variation and are therefore susceptible to natural selection. Sometimes the selection on a particular trait will be directional, but for most of the individual traits for most of the time, stabilizing selection occurs in order to maintain the functional integrity of the phenotype as a whole. The second implication is that selection acts on the phenotype as a whole, as measured by its reproductive fitness, and therefore it acts on the integrated outcome of all the individual traits. The contribution to the overall phenotypic fitness of any one particular trait at any one moment in evolutionary time may be high, low, zero, or even negative if that trait nevertheless happens by chance to occur in the fittest phenotype. In fact, the fitness contribution of a trait depends on a combination of the particular environmental features to which it is relevant, and on how well it integrates functionally with all the other traits. For example, suppose a mutation arises which causes an increase in the oxygen-carrying capacity of the blood. If oxygen supply to the tissues is limiting, then the mutation will be positively selected. However, if the increased oxygen tension in the tissues also adversely upsets, say, the acid-base balance, then on balance the mutation might be neutral. If the deleterious effect was greater than the beneficial effect, the mutation considered in isolation would be deleterious, but not necessarily to an extent that outweighed a high level of fitness of the organism as a whole resulting from the interaction of all the traits.

The principal feature of an adaptive landscape that illustrates the implications of correlated progression is very high dimensionality in the sense that each of the large number of genetically variable traits has to be represented by its own axis in morphospace. The effect of the simultaneous selection of many traits is to reduce the evolvability of any one specified focal trait, even if that trait is facing directional selection (e.g. Hansen 2012). The need to maintain its functional relationship with many other traits holds it in check, as it were. One version of an adaptive landscape that approaches these requirements (Fig. 6.1a) is the 'holey' landscape of Gavrilets (1997), although his actual formulation concerns gene combinations at the population genetics level. He showed that with a very large number of dimensions, the topographical map representing the fitness surface will be close to flat, which is to say there are many different genotypes that are of more or less equal fitness. However certain gene combinations are assumed to have a lower fitness, and he represented these by areas of the flat fitness surface than cannot be occupied. They are described as 'holes' and it is assumed that they represent not just gene combinations that are less fit, but that are biologically inviable and therefore are parts of the landscape that cannot be entered or crossed by an evolving lineage. Under these specified conditions, the evolution of new combinations of genes occurs largely by neutral drift around the plane surface, while avoiding the holes. Translating the holey adaptive landscape into a phenotypic context means first that there are many potentially evolvable traits, and second that many combinations of their values constitute more or less equally well-adapted organisms. This is consistent with the assumption of the correlated progression model that there is enough flexibility (tolerance of minor

variation) in the functional linkages between integrated traits to allow a wide range of sequences of successive small changes in many traits to occur with no significant change in overall fitness. As in Gavrilets' version of the holey landscape, this permits a lineage to evolve around the fitness plane more or less at random, by accumulating small changes in many of the traits, the possible changes being dictated mainly by the requirement that all traits retain functional integrity with each other. It is also compatible with Haldane's cost of selection (see section *The implications of the modularity model on evolution*, Chapter 3), since many of the changes in traits can be selectively neutral.

This phenotypic modification of the holey adaptive landscape captures certain features of the correlated progression model and may offer a satisfactory metaphor for radiations at low taxonomic levels, but it does not illuminate evolutionary transitions to new higher taxonomic levels. In Gavrilets' adaptive landscape, provision has to be included for evolution to a new adaptive level and it takes the form of occasional steep slopes, or 'ramps' (Fig. 6.1b) up which selection could drive a population (Goodnight 2012). Thus, while the model allows for neutral drift in trait values leading to low taxonomic-level demes and perhaps species, for a more significant transition between higher-level taxa, a ramp must be ascended. A ramp climbing steeply to a new adaptive plane represents the situation where a relatively large evolutionary change results in adaptation to a significantly different environment.

If it were the case that a ramp rises in a single or small number of dimensions, climbing the ramp would be achieved by large change in a single or small number of traits, which corresponds to a frequently used model for major transitions based on the concept of a key adaptation. A key adaptation, or key innovation, is a presumed change in a single character, or at most a biological property of the phenotype, that permits entry into a new adaptive zone, or adoption of a new mode of life (Hunter 1998; Galis 2001). Originally introduced by Simpson (1944), key innovations are mostly discussed in the context of adaptive radiations at a relatively low taxonomic level, but sometimes are assumed

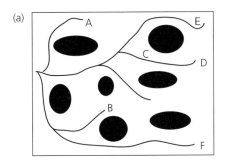

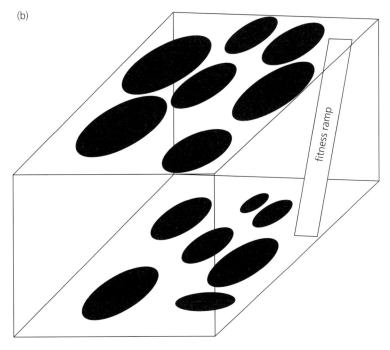

Figure 6.1 (a) A 'holey' adaptive landscape. Lineages diverged and taxa A to F arose by near-random walking on a flat adaptive landscape with little difference in fitness, except that certain trait combinations, represented by the black holes, are highly unfit and cannot evolve. (b) Gavrilet's concept of a ramp leading to a new higher taxon which itself occupies a holey adaptive landscape. From Goodnight 2012, reproduced with permission from Oxford University Press.

to initiate the origin of higher taxa. Examples of putative key innovations are as diverse as the pharyngeal teeth of cichlid fish (Wainwright et al. 2012) and the hypocone of mammalian molar teeth (Hunter and Jernvall 1995). The difficulty with the concept of key innovations for higher-level taxa is that it involves abandoning the correlated progression model along with the arguments in its favour altogether. The emergence of a higher taxon involves changes in too many and too diverse a range of integrated traits to allow any one of them to be singled out as paramount to the evolutionary transition, as witnessed by the fact that there is never a major innovation in a single trait defining or leading towards a new higher taxon. Any putative candidate for the status of key innovation associated with the origin of a higher taxon is always accompanied by numerous evolutionary changes in other traits, and there is no logical basis for regarding any one of them as more 'key' to the transition than others. Therefore, to be consistent with correlated progression, the ramp in the adaptive landscape leading to a new higher taxon must be envisaged as rising up in many dimensions, which is to say climbing it requires correlated changes in many characters. Yet if this is the case, then the essential flatness of a highly dimensional adaptive landscape is not escaped. Therefore the question for understanding the ecological conditions for major evolutionary transitions can be expressed as how the lineage can evolve in a flat fitness landscape involving large numbers of traits under simultaneous selection and yet evolve in a directional manner leading from an ancestral to a radically new, derived higher taxon. What environmental circumstances eventually drive a simple worm-like animal to end up as an octopus or lobster; a squat, sprawling, scaly, ectothermic, tiny-brained reptile as a bird or a mammal?

6.2 Multidimensional gradients in the adaptive landscape

The long-term trends shown by evolving lineages that achieve the status of new higher taxa by changing numerous characters over time are therefore unconvincingly accounted for by low taxonomic-level phenotypic wandering over an essentially flat fitness surface. They attest rather to a more deterministic process in which there are features of the adaptive landscape that they can, as it were, track. Since under the conditions of the correlated progression model only small correlated changes in characters accumulating gradually are permitted, the implication is that this tracking by a lineage consists of following an ecological gradient in which changes in the environmental parameters encountered are also gradual. However, since many characters do change, affecting many parts and functions of the phenotype, the ecological gradient itself must be multidimensional. Consider, for example, a hypothetical species in a niche defined by particular food type, locomotory requirements, ambient temperature and humidity range, and

guild of predators. For each of these environmental variables there is an ecological gradient leading to a different condition. There must therefore also be compound gradients in which all the variables change together: say in the direction of larger food particles, more active locomotion, greater ambient temperature range, lower humidity, and mean predator size. The effect on the evolving lineage of such a gradient will be that, of all the viable combinations of trait changes, there will be one particular suite of changes that in total would be an adaptive response to it. In the example, evolution of the feeding and locomotory systems, the physiological adaptations for seasonality, and the behavioural responses to predation would all occur. With many traits involved that are associated with the different respective functions of the organism, coupled with the requirement to maintain functional integration amongst them, tracking such a compound gradient is explicable by the correlated progression model.

Ecological gradients compounded of several parameters of the habitat as so envisaged offer an answer to the question of what drives a lineage to the point of becoming a new higher taxon. The gradient can be represented as a flat-topped ridge in a multidimensional adaptive landscape (Fig. 6.2). Its narrowness represents the relatively constrained combination of trait changes necessary to track all the gradually changing environmental variables at once: a large change in any single trait would lead down the side of the ridge to a less overall fit state. The flat top of the ridge represents the degree of flexibility of the functional relationships of the traits demanded by the correlated progression model, and allows for the minor radiations at various stages as the lineage tracks along it.

Other features of the origin of new higher taxa can be represented by adaptive ridges of this nature. As a percentage of all the diverging lineages over evolutionary time, the number that ended up as new higher taxa such as orders and classes is exceedingly small, while the vast majority terminated only as low-level taxa such as species and genera. This suggests that compound ecological gradients of sufficient complexity or number of dimensions to be associated with the emergence of a new higher taxon are rare. The ridges must also be long

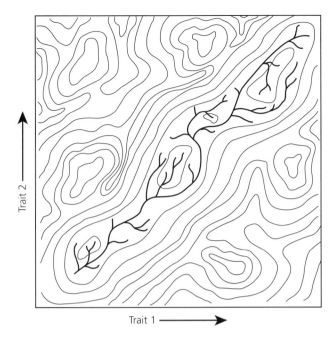

Trait 2

Trait 1 ⟶

Figure 6.2 A more or less flat ridge in an adaptive landscape and an evolving lineage tracking it. Correlated changes in Trait 1, Trait 2 . . . Trait *n* must all occur.

enough—that is to say associated with a sufficiently large amount of evolutionary change—to qualify as leading to a new higher taxon; in terms of another metaphor, they must traverse large distances through morphospace.

The ridges must also be relatively long-lived in evolutionary time, since they are supposedly responsible for lineages tracking them over millions to tens of millions of years. However, a particularly important property of the evolving lineages is that the sequence of ancestral grades and their low-level radiations, with only occasional exceptions, become extinct. This can be represented as a continual loss of the ridge as the lineage passes along, corresponding to ever-changing environmental conditions. The main change includes the presence of the new, more derived stage itself. However, to be more realistic, adaptive landscapes must always be regarded as constantly changing in topology over time (Simpson 1944; Arnold et al. 2001; Aguilée et al. 2012; Hansen 2012). New adaptive possibilities and constraints on evolutionary change continually arise due to evolution of the community coupled with constant modifications to the abiotic environment. Indeed, speeded up over time, an evolving lineage might better be regarded as traversing an adaptive

sea of waves and currents rather than a stable landscape. It is certainly not necessary to assume that the hypothesized ecological gradients existed in their entirety from the start to the completion of the evolutionary transition. As well as imagining parts of it disappearing behind the lineage as it moves on, so also extensions may arise in front of the lineage as it approaches.

With imagination therefore, it is simple to invent an ecological model in which from time to time new higher taxa evolve, and ancestral ones become extinct, but whether such a situation as is represented by these adaptive ridges actually occurs in the real world is difficult to establish. However, it is possible to assess how far this version of an adaptive landscape is consistent with those rather few cases where the pattern of acquisition of derived characters is adequately known from the fossil record.

6.3 Case studies: aquatic to terrestrial habitats

Several major new taxa are associated with a transition from an aquatic to a fully dry land habitat, notably insects, arachnids, myriapods, and tetrapods.

Comparing the two habitats, there are several physical and chemical differences between water and land that are often referred to as 'problems' to be solved by prospective terrestrial organisms. These are gravity due to the loss of buoyancy causing high body weight to be supported and moved, desiccation due to evaporation in air, and high ambient temperature fluctuation inimical to regulated biochemical activity. There is also a requirement to shift from aquatic to aerial respiration; from detection of waterborne sound and chemo-stimuli to sensing airborne modalities; and from external to internal fertilization. In the case of the vertebrates, a shift from the suction-feeding orobranchial cavity of typical fish to the grasping jaws characteristic of terrestrial vertebrates must be added. Looked at purely physically, water and land appear to separate two entirely distinct modes of life, and the water–land interface to be a particularly severe barrier for an evolving lineage to cross. However, in terms of the necessary biological adaptations, an ecological gradient or continuum between the two can be described (Fig. 6.3).

The case of the terrestrial tetrapods is the best understood because there are a number of intermediate grades in the fossil record, as described in Chapter 8, section 8.3.1. Typical fish-like locomotion depends on the buoyancy of the water and the absence of obstructions to permitting movement by the lateral undulation of the body and the median and caudal fins. The lobe fins of the tetrapodiform 'fish' contained bones and musculature for an alternative mode of locomotion in which the rigidity and strength of the appendages could propel the body by direct contact with the substrate. This is suitable for densely vegetated and shallow bodies of water in which motion of flexible planar appendages is less effective. Increasing the size, robustness, and mobility of the appendages relates to increasing weight-supporting ability, and therefore successive stages of the evolving lineage spent a gradually increasing percentage of time partially and eventually fully exposed to the air. A gradient in the mechanical parameter between water and land can therefore be envisaged, along which increasingly robust rather than flexible appendages represent the adaptive response. A gradient of feeding function can also be described. The shallower the water, the greater is the percentage of food items that must be picked off the surface or off the nearby bank, and therefore the more important the grasping rather than sucking mode of intake. Adaptations of the physiology and biochemistry from a condition suitable for complete submergence in deep water, through shallow water, to partial exposure to air, and eventually to full exposure to air, represent an evolutionary response of a lineage over time to a gradient of increasing potential desiccation and magnitude of daily and seasonal temperature fluctuation. Specific adaptations here include a complex of features of the external integument, chemo-regulation system, and circulatory system along with appropriate, neurally controlled

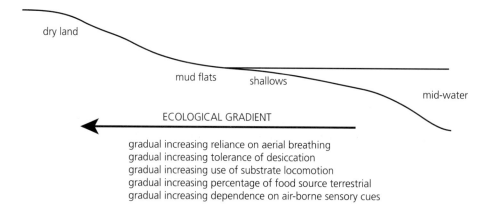

Figure 6.3 The multiparameter ecological gradient between aquatic and terrestrial life tracked during the transition from fish-grade to tetrapod-grade vertebrates.

behavioural responses. To judge from modern air-breathing fish, aerial breathing evolved initially as a facultative adjunct to, not a replacement for, aquatic respiration, and through the sequence of phenotypes the lungs evolved an increasing capacity and the gills a decreasing one until finally the latter were lost, again a situation well-described as an adaptive response to a gradient of increasing reliance on atmospheric gas exchange. In a comparable fashion, reception of airborne sensory modalities can gradually replace aquatic ones as the percentage of each category to which the animal was exposed changed. An ecological gradient associated with developing eggs can be envisaged as a gradual shift to eggs increasingly resistant to desiccation being laid in decreasingly less humid conditions. The actual details of the evolution of a fully terrestrial egg are not understood, nor how phylogenetically the modern amphibians with their aquatic eggs and larvae fit in, but at least on the line to amniotes it began as eggs resistant enough and well enough supplied with yolk to develop under moist terrestrial conditions (Martin and Carter 2013). The transition from fish grade to tetrapod grade can therefore be postulated as evolution along a complex, multidimensional gradient between water and land, and therefore as the tracking of an adaptive ridge by the lineage.

The actual fossil record of the origin of tetrapods reveals a pattern of acquisition of tetrapod characters consistent with the correlated progression model (Chapter 8, section 8.3.2). The reconstructed sequence from a fish grade as represented by *Eusthenopteron* to a basal tetrapod with limbs, loss of the gills, etc., shows that modifications to practically all parts of the skeleton and, where inferable, also to soft tissues and physiological functions occurred in concert. These include adaptations associated with locomotion, aerial breathing (Clack 2006, 2012), grasping rather than suction feeding (Markey and Marshall 2007), and for detecting airborne sound and olfactory modalities. There is little direct evidence of the extent to which there were also adaptations for withstanding increased ambient temperature fluctuations, evaporative water loss, and terrestrial reproductive behaviour, although given the presumed habitats involved they may be confidently inferred. Assuming that these various

evolving functions were indeed driven by natural selection, then the simultaneous responses to several different environmental variables supports the idea of a complex ecological gradient leading from a fully aquatic ancestral stage to a terrestrially adapted tetrapod grade. The various functions may not have evolved at the same rate as each other throughout the process, and at any one moment in evolutionary time some were perhaps ahead of others. For example, Anderson and colleagues (2013) produced evidence that the mandible structure evolved more slowly; however there is no evidence for any particular one of the functions having evolved to any great extent without major shifts in others.

The morphological evidence of the fossils therefore supports the idea of a high-dimensional adaptive gradient existing between the two habitats. It ran from mid-water, to increasingly shallow water, to an amphibious life on the muddy banks, to terrestrial life without any significant break. Furthermore, each of these successive stages itself represents a viable habitat within which a low taxonomic-level radiation can occur, although at present insufficient fossil evidence is available to describe these in any detail.

The fossil record of the transition from fish grade to tetrapod grade occurs within the Devonian period. The earliest body fossils of the tetrapodiform lineage are Frasnian in age, which dates them between 385–375 Ma, while definitive tetrapods with limbs are present by the Famennian, which is 375–360 Ma. The picture is, however, complicated by the discovery of tetrapod footprints from sediments in Poland that are dated to the Middle Devonian, at about 395 Ma, which implies a considerably earlier origin of tetrapods than formerly supposed (Niedźwiedzki et al. 2010). For many years the transition was believed to have occurred in a freshwater setting because the only known Devonian tetrapods were *Ichthyostega* and *Acanthostega* from the continental rocks of East Greenland. However, several subsequently described taxa are from tidal, marginal marine, and estuarine environments. The Polish footprints, representative of the earliest known tetrapods, occur in a coastal area of shallow marine lagoons separated by low-relief land (Narkiewicz and Retallack 2014). The fossils are also mostly from deposits laid down within regions that at the time were within the northern hemisphere tropics where global temperatures were around the mid-30s, and so somewhat higher than today's.

The ecological conditions of this period are broadly understood. Most noteworthy is the estimate of the percentage of oxygen in the atmosphere. Using a variety of lines of evidence and computer modelling, Berner and colleagues (Berner et al. 2007, 2009) estimated that there was a rise to about 25 per cent during the Early Devonian but this had fallen to about 17 per cent at the end of the Frasnian and then up again to about 22 per cent by the end of the Devonian. Attempts to relate this to the origin of the tetrapods are ambiguous. Clack (2006, 2012) argued that the anatomy of the spiracular region of the back of the skull indicates an increased reliance on air-breathing, correlated with the fall in oxygen levels during the Frasnian. George and Bliek (2011), by contrast, proposed that the origin of tetrapods was associated with the high oxygen levels of the Early Devonian. It is oversimplifying the situation to consider a single parameter to be a primary cause of such a major evolutionary transition, for there were great changes occurring in the biota as a whole. From the Silurian and throughout the Devonian, the diversity and average size of near-shore and terrestrial plants was increasing, and along with this the diversity of terrestrial arthropods such as millipedes, several arachnid taxa including the large-bodied trigonotarbids, and hexapods.

A series of interrelated parallel ecological changes can therefore be seen during the course of the Devonian. Increasing plant density and diversity produced heavily congested shallow-water habitats along the continental margins, estuaries, and freshwater systems for which substrate locomotion culminating in replacement of fins by limbs was the adaptive response. Simultaneously, the oxygen levels gradually fell thanks in part to the relatively high ambient temperature, perhaps in part to increased run-off of nutrients from the terrestrial plant life causing algal blooms (Algeo et al. 2001). Meanwhile, the increasing terrestrial arthropod biota must have provided a gradually increasing new source of food for which capture by grasping with jaws rather than suction within water was achieved by evolution of a new jaw mechanism,

and increasing reliance on airborne sensory modalities for its detection. Along with this latter benefit of terrestriality comes the need to adapt to increasingly extended times of exposure to high and fluctuating ambient temperatures.

A scenario such as this offers an account of the cause of the origin of tetrapods that avoids the unrealistic assumption that any one ecological parameter was sufficient to cause the entire transition. It also allows for the radiations that the fossil record at least hints occurred at various levels or grades along the principal lineage. Most importantly, it is compatible with the pattern of acquisition of tetrapod traits inferred from their cladistic relationships, specifically with the correlated progression model.

None of the cases of invasion of the terrestrial habitat by invertebrate taxa are adequately illustrated by fossil evidence, although it may be speculated that they followed broadly the same ecological gradient as the tetrapods. The main difference in the case of the insects was that initially the new source of food for which they became adapted consisted of terrestrial plants.

6.4 Case studies: low-energy to high-energy life styles

Two modern higher taxa, mammals and birds, evolved endothermy which was a revolutionary new biological organization based on relatively hugely increased metabolic rates, typically some six to ten times those of ectotherms. Of the two, the fossil record of the lineage from basal amniote to mammals is the most complete example of the origin of a major new taxon. As discussed at length later (Chapter 8, section 8.1.3), it reveals a pattern of acquisition of mammalian characters that is completely compatible with the correlated progression model. Features of the skull indicate that changes in the dentition and jaw musculature led to more effective feeding. A secondary palate separated mastication from breathing, and the form of the ribcage indicates the evolution of a diaphragm and greater lung capacity. Functional analysis of the postcranial skeleton demonstrates that locomotion was more effective and manoeuvrable. There is also evidence that both hearing and olfactory sensory reception became more sensitive and that the brain increased in volume. The bone histology indicates indirectly that growth rate was increased. Taken altogether, the evidence points overall to increasing metabolic rate and activity level along the lineage. Furthermore, these numerous biological changes evolved hand in hand with one another over the 100 million years or so that it took the lineage to evolve a fully endothermic mammal from the basalmost ectothermic stem member.

The biological evolution along this lineage therefore involved correlated adaptive shifts in response to several aspects of the environment, both abiotic such as rate of gas exchange and operating in a complex physical terrain, and biotic such as the rate of acquisition and assimilation of food, and sensitivity to sensory cues. This multidimensional adaptive response can be expressed as the lineage tracking a ridge in an adaptive landscape in which the various parameters relate to increasing metabolic rates (Fig. 6.4). Several functional consequences flow from the high-energy, endothermic mode of life. First, accurate regulation of the internal body temperature in the face of a fluctuating ambient temperature depends on high cellular energy levels, as does the maintenance of a constant internal chemo-environment when exposed to dry and arid conditions. As dimensions of the adaptive ridge, these related to the ability to remain active over a wider range of ambient temperature and humidity levels, such as occurs on both a diurnal and a seasonal time scale. Second, a consequence of a high basal metabolic rate is that a high level of aerobic activity level can be achieved and sustained, which translates to a variety of strategies of food capture and predator avoidance. Third, thanks to the accurate regulation of body temperature and molecular composition available to an endotherm, a larger, more complex brain is possible which can accept, integrate, and make use of more information. This allows more precise control of muscular action such as in locomotion in a physically heterogeneous habitat and complex foraging strategies, and greater adaptability of behaviour. Taken together, the evolving lineage that ultimately led to mammals tracked an environmental gradient or adaptive ridge of increasing independence of ambient conditions daily and seasonally, and increasing adaptability of behaviour such as in food acquisition, predator avoidance, and presumably sociality.

The fossil record of the mammalian lineage also shows a series of relatively short-lived radiations of a number of grades. These can also be expressed in terms of the adaptive landscape, as low taxonomic-level variation around a particular section of the ridge superimposed on the general trend.

The direct palaeoenvironmental evidence is also consistent with the hypothesis. The basalmost grade are the pelycosaurs and they are almost exclusively restricted to continuously humid equatorial environments of the Lower Permian, implying thermal physiology akin to modern ectotherms, and limited osmoregulatory ability (Kemp 2006). They appear to have relied on the continual presence of free water, and this interpretation explains a particular anomaly of pelycosaur communities, namely that there was a relatively very high ratio of carnivorous forms to herbivores. At this time the terrestrial community evidently still depended on a good deal of primary productivity and primary consumption by aquatic organisms. Transfer of this energy to terrestrial tetrapods occurred via semi-aquatic amphibians and fish-eating taxa. In contrast to the pelycosaur-grade dominated communities of the Lower Permian, the next main grade occurred in temperate latitudes during the Middle Permian. Here the climate was seasonal, consisting of warm, dry periods alternating with cooler, wetter periods and which indicates that these, the basal therapsids, remained active over a greater range of environmental temperatures and humidity levels. Apart from a very small number of species, pelycosaur-grade synapsids had disappeared by this time.

The next environmental indicator of an increased regulatory ability is perhaps the survival of the end-Permian mass extinction. The trigger for this event and the accompanying loss of the great majority of Upper Permian taxa was almost certainly massive volcanic activity associated with the huge area and depth of volcanic rocks, the Siberian Traps, of northern Asia (Twitchett 2006). Global warming, thanks to CO_2 output and the release of hydrates, increased climatic temperatures, whilst SO_2 output caused serious depletion of terrestrial plant life. The effect was an increase in seasonal aridity as well as average global temperature, and presumably the

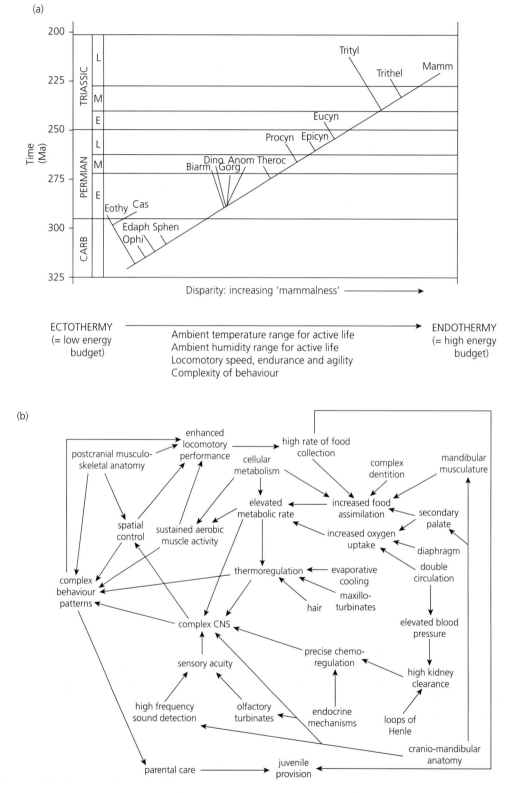

Figure 6.4 (a) The parameters of the adaptive gradient, or ridge leading from a low-energy ectothermic mode of life to a high-energy endothermic mode in mammals. The cladogram shows the sequence of fossil taxa occupying different parts of the gradient over time. (b) The integration of the many structures and processes that had to change as the lineage tracked the adaptive ridge. (After Kemp 2006.)

small number of survivors amongst the therapsids, including the first of the cynodonts that constituted the next main radiation, were adapted to cope. Notably, these had evolved several mammalian traits associated with increased metabolic rates and activity levels. More progressive cynodonts radiated throughout the Middle and Upper Triassic, culminating in the earliest mammals. These had a greatly reduced body size, mostly no larger than modern rodents. Much has been made of the implication of miniaturization for their biology, and in particular that they were nocturnal, feeding on a high-energy diet of insects. To achieve this habitat requires above all regulation of the body temperature, allowing continual activity under the cooler conditions of night time.

To conclude, this sequence of environmental circumstances together—equatorial humid, higher latitude seasonal, highly arid seasonal, and eventually nocturnal—is consistent with the proposed long-term evolutionary lineage tracking of a general environmental gradient associated with lifestyles demanding increasing metabolic rates, internal regulation, and activity levels. In turn, the scenario is consistent with the correlated evolution of characters associated with several biological functions in concert.

There is a good deal less detailed anatomical evidence from the fossil record concerning the origin of the birds from basal archosaurs (Chapter 8, section 8.2.1), although what there is points to a comparable situation to that of mammals. The basalmost archosaurs undoubtedly had low metabolic rates, and in the course of the lineage culminating in birds, many characters associated with endothermy and the associated sophistication of locomotory, feeding, respiratory, and sensory systems evolved in concert. Therefore a similar adaptive ridge may be postulated for this taxon, based on the acquisition of adaptations for increasing endothermy, and thus the evolution of increasing independence of ambient environmental fluctuations, high-energy activity to the extent that active flight became possible, and complex behavioural patterns.

6.5 Case studies: the origin of modern invertebrate body plans

The fossil record offers a view at once intriguing and frustrating into the environmental circumstances of the origin of the major invertebrate taxa. As described in detail elsewhere (Chapter 7, section 7.2), the earliest fossil evidence for animal life is the Ediacaran fauna, dated from the eponymous stage of the late Proterozoic. Virtually all specimens are impressions of relatively large soft-bodied organisms and their affinities continue to be matters of considerable disagreement (Laflamme et al. 2012). At one extreme is the view that they constituted a distinct grade of animal life based on flat surfaces rather than compact bodies. As such they have been referred to as the Vendobionta (Seilacher 2007). At the other extreme, several of them have been interpreted as stem members of later metazoan phyla, such as sponges, cnidarians, annelids, molluscs, arthropods, and echinoderms. They have even been considered as primitive fungi or plants (Retallack 2013).

Undoubted stem members of modern metazoan phyla are not represented in the fossil record until a relatively narrow window of time in the Lower Cambrian, although Budd and Jensen's (2000) warning must be borne in mind not to equate a fossil that has one or a few characteristics of a phylum as evidence that the complete body plan of the crown group had evolved. Careful analysis of the relative dating of the fossiliferous strata has shown that this event is not quite the 'explosion' that it is normally described as being, but nevertheless it occurred rapidly (Maloof et al. 2010). The first manifestation is in the earliest Cambrian stage, the Nemakit-Daldynian (540–525 Ma). This is the beginning of the 'small shelly fauna' (Fig. 7.4) which comprises the first biomineralized animals. Along with numerous fossils of unknown affinities, there are specimens identified with varying degrees of confidence as sponges, cnidarians, univalve and bivalve molluscs, brachiopods, and possible fragments of arthropods. Larger members of these groups and also trilobites are preserved in the succeeding Tommotian stage (525–521 Ma), while the Atdabanian-stage deposits from Chengjiang in China and Sirius Passet in Greenland reveal a great variety of phylum-level groups. Thus, as far as the fossil evidence is concerned, the Cambrian explosion occurred over the course of about 20 million years.

Molecular-based divergence dates for the origin of phyla must be treated with some caution, and the distinction maintained between the divergence of two lineages which is what the molecular clock indicates, and the origin of recognizable new body plans, which is what morphology indicates. While ambiguities associated with the actual phylogenetic relationships amongst the phyla and superphyla are largely resolved (see Fig. 7.2), the constancy or otherwise, and the calibration of the molecular clock used to estimate the age of modern, crown taxa remain problematic. Unsurprisingly, estimates are therefore very variable. At one extreme are estimates of divergences extending well back before the Ediacaran (Hedges and Kumar 2009): over 1000 Ma for superphyla and 500–800 Ma for phyla. On the other hand, a number of other authors have estimated dates for the origin of taxa that are much closer to the appearance of the earliest members in the fossil record (Peterson et al. 2008; Edgecombe et al. 2011; Erwin and Valentine 2013), with the origin of the superphyla in the Ediacaran, and the phyla late in the Ediacaran or early in the Cambrian. If the older dates are correct it implies that there was a very long period during which no fossils were formed, perhaps because the lineages were represented by small, soft-bodied forms. If the more recent dates are correct, it implies that there was a very rapid evolutionary transformation of ancestral-grade forms into recognizable members of the modern phyla. Either way, there was something in the Early Cambrian environment conducive to the origin of a series of new body plans.

There is evidence of radical changes in the environmental conditions over this period of time, and several aspects of the later Neoproterozoic world seem likely to be relevant to the rise of the phyla. One is a series of glaciation periods referred to as the 'Snowball earth'. Oxygen isotopes and other geochemical signals indicate that there were several ice ages during this time, two of which were particularly severe with the ice sheets extending right into the equatorial regions and therefore virtually covering the earth. The cause of these episodes is not clear, but associated shifts in carbon isotope ratios indicate changes in the carbon cycle involving the balance between atmospheric CO_2, organic carbon locked up by photosynthesis, and carbon returned to the sea to be locked up as inorganic carbonate (Hoffman et al. 1998; Tziperman et al. 2011). A second major abiotic change was a rise in deep oceanic oxygen levels. The increase in oxygen levels between about 800–540 Ma is well documented by several geochemical techniques (Canfield et al. 2007; Frei et al. 2009). Estimates of the level of oxygen vary between about 15 per cent and 40 per cent of the present-day level by the late Ediacaran, which is high enough to sustain active aerobic multicellular life (Butterfield 2009a), and indeed the commonest explanation given for the Cambrian explosion is that this rise in oxygen level was the trigger because it allowed increased metabolic rates and therefore body size. A third abiotic environmental shift was some threefold increase in the calcium content of the oceans. This probably occurred due to a large Ediacaran–Cambrian transgression in sea level causing increased weathering of the continental soils and rocks (Peters and Gaines 2012), and could have been responsible for the parallel evolution of calcified hard skeletons in several phyla over this period (Murdock and Donaghue 2011), perhaps initially as an adaptation for sequestering otherwise toxic levels of calcium ions in the body tissues.

It is clearly far too simple to suppose that any of these abiotic changes alone or even in combination constitutes a sufficient explanation of the Cambrian explosion; at best they may have been triggers, and may even have been consequences rather than causes of the evolutionary events. This is clearly the case for oxygen levels which are determined in large part by the balance between aerobic photosynthesis and oxidation of carbon. Tziperman and colleagues (2011) proposed that the Late Proterozoic glaciations resulted from a fall in atmospheric carbon dioxide following the evolution of larger planktonic organisms, because these more rapidly sank to deep, anoxic ocean layers taking carbon with them. Thus it has been stressed that the very rise of new kinds of organisms may itself have been an integral part of what drove the Cambrian explosion and emergence of the animal phyla. Butterfield (2009, 2011) has developed a scenario for the transition from a pre-Ediacaran world in which the evolution of Eumetazoa was itself a significant cause. Pre-Ediacaran communities were dominated for some 2 billion years by microorganisms in which dense accumulations of Cynanobacteria with very low sinking rates created turbid, anoxic conditions inimical to aeroboic metazoans. The evolution of suspension-feeding eukaryotic zooplankton that fed on the microorganisms, was the adaptive force behind the evolution of larger bodied phytoplankton with much higher sinking rates: this resulted in better-oxygenated, clearer water conditions. From this point, diversification of eumetazoans with the potential to evolve increased body size led to a range of different ecotypes: faunivores with specialized mouthparts and planktivores with protective devices, each kind acting as a selective force for others in an arms race with multiple participants. Of course, the origin of the phyla also involved the origin of their molecular developmental mechanisms. As well as the appropriate environmental circumstances and biotic interactions providing the selection forces around the end of the Neoproterozoic, these potential mechanisms must have been largely in place by this time (Butterfield 2009; Marshall and Valentine 2010; Erwin et al. 2011). This aspect was considered in Chapter 5 and in the course of summarizing the ecological picture here it is taken for granted.

The idea that the origin of a new higher taxon can be envisaged as a lineage tracking a flat-topped ridge across a multidimensional adaptive landscape can be applied well to the present case of the animal phyla. Oxygen and calcium levels in the seawater were increasing through the Ediacaran and early Cambrian, and therefore presented a geochemical gradient through time to which an evolving lineage could respond adaptively. A little more subtly, the novel biotic interactions that were arising at this time as a consequence of increasing metabolic rates and calcification may also be described as gradational rather than incremental over time. Increasing size and activity levels of a lineage of predators presents a lineage of non-predators with a gradual parameter requiring a gradual adaptive response. Conversely of course, the predators must co-evolve with these avoidance strategies developing in their prey. The special aspect of the Cambrian explosion lies not in the principle of co-evolution amongst community members as such, but in the novelty of the high oxygen and high calcium environment which had never previously existed. Ecological opportunities for diversification had never been so abundant. Nor were they ever to be so abundant again, once they had been grasped and refined by the end of the Cambrian.

The multiplicity of lineages each tracking the same broad adaptive ridge but leading to different phyla meets the expectations of the correlated progression model, and indeed is difficult to account for otherwise. Adaptive response to the several abiotic and biotic parameters of the gradient requires integrated changes in numerous phenotypic traits. An increasing metabolic rate is permitted by the increasing oxygen level, which in turn allows increasing body size. But as size increases, so skeletal support, specialized respiratory organs, active circulation of body fluids, and increased rates of food collection are increasingly necessary. For those lineages that evolved a calcified skeleton, the physiological processes of calcium metabolism need modifying. As predator avoidance and prey capture behaviour evolves, so locomotory organs and their performance, protective devices, sensory sensitivity, and central nervous system control must all modify. For capture of prey of increasing size, more specialized mouthparts must evolve. For functional integration of the whole organism, the central nervous system needs to evolve. All these structures and functions, each itself divisible into several integrated traits, must evolve in a correlated pattern. Under the precepts of the correlated progression model, traits evolve by small increments, maintaining integration over time, and the exact order and pattern of trait changes in any lineage depends on the precise adaptive exigencies of the moment, the relative degrees of tightness of integration amongst specific traits, and the kind of variation available in the traits. Inevitably, different lineages evolve by different patterns of acquisition of new traits, leading to different end products.

CHAPTER 7

The invertebrate fossil record

The fossil record of stem members of crown invertebrate phyla is sparse and consists almost entirely of Cambrian specimens, of which much the most important are found in an extraordinarily finely preserved condition in a number of Lagerstätten. These are deposits which for various fortuitous reasons, such as certain geochemical conditions or very rapid anoxic burial, preserve the organic carbon of the tissues as well as the normal calcareous exoskeletons. For all the detailed, often beautiful anatomy of the specimens, the actual structure is often very difficult to interpret. In many cases the organisms are very unlike any living forms, and their characters are often difficult to identify because they lack enough structural detail to be sure what their homologues are in modern taxa, even to the extent that in several cases there are arguments about to which, if any, living phylum a fossil belongs. Even in those cases where the phylum membership is not seriously in doubt, these fossil taxa tend to possess either very few of the derived characters of their phylum, or else almost all of them. Either way, they shed relatively little helpful light on the pattern and sequence of acquisition of characters of the crown group. In fact most of the inference about the morphological evolution of the phyla is based on comparative and embryological evidence of modern forms rather than on direct palaeontological evidence of sequences of stem taxa. Consequently little is actually known about how the characters and body plans of higher invertebrate taxa were assembled.

7.1 The phylogenetic tree of crown invertebrate phyla

As recently as, say, Willmer's (Willmer 1990) textbook, there were conflicting views even about the principal lineages of invertebrates, while the interrelationships at the phylum level were largely an unresolved mystery (Fig. 7.1a). Indeed, her concluding 'approximation to an invertebrate phylogeny' consisted of about 24 separate lineages leading away from an amorphous centre (Willmer 1990, Figure 14.2). The only two monophyletic groupings of much substance included were the Deuterostomia containing chordates, echinoderms, and hemichordates, and an unnamed taxon containing annelids, hexapod and myriapod (but not crustacean or chelicerate) arthropods, onychophorans, pogonophorans, and echiurans.

Although criticized by some for its non-cladistic approach, Willmer's conclusion is actually an accurate reflection of the extreme lack of agreed resolution of the phylogeny at that time. Other authors passionately argued amongst themselves for one or another claimed monophyletic groupings, but with little consensus. It was assumed that the problem lies in the very long time that has elapsed since the original diversification into the stems of the modern phyla, possibly coupled with a relatively short period of time over which this diversification actually occurred. In modern terminology, the internal branches of the invertebrate tree are short while the external branches are long. This increases greatly the difficulty of discerning homology amongst greatly modified characters, and of recognizing the polarity of character changes. It also leads to the problem of long branch attraction, in which the greater the amount of evolutionary divergence between lineages that has occurred, the greater the probability of two or more of them evolving similarities by convergent evolution. In fact, the majority of characters regarded as informative are embryological or larval, which are highly variable even within

The Origin of Higher Taxa, First Edition. T. S. Kemp.
© T. S. Kemp 2016. Published 2016 by Oxford University Press and The University of Chicago Press.

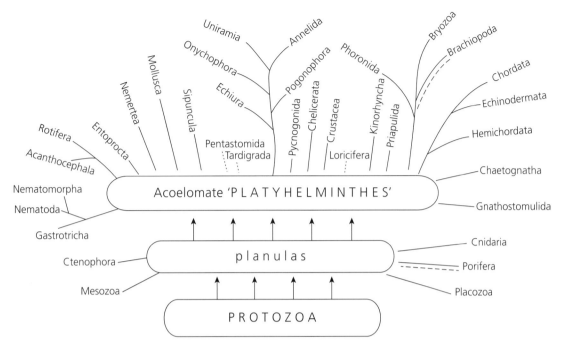

Figure 7.1 Wilmer's pre-molecular, non-cladistic phylogeny of the invertebrates. (From Wilmer 1990, reproduced with permission from Cambridge University Press.)

undisputed monophyletic taxa. Thus, for example, it was not agreed whether molluscs evolved from a segmented, worm-like common ancestor shared with annelids, or an unsegmented, muscular-footed ancestor shared with platyhelminthes. Even the monophyly or polyphyly of Arthopoda remained a matter of dispute.

From the mid-1970s, molecular sequence data started to become available, and molecular-based phylogenetic relationships to be proposed, including some most unexpected ones (Fig. 7.2). First, on the basis of ribosomal RNA Halanych and his colleagues (Halanych et al. 1995) found a monophyletic group consisting of the annelids, the molluscs, and the lophophore-bearing brachiopods, phoronid worms and bryozoans, a group that was later named Lophotrochozoa. Shortly afterwards, and equally radically, Aguinaldo and colleagues (1997) proposed a relationship amongst the phyla which moult their cuticle. Now known as Ecdysozoa, it included all the arthropods, the nematode and the priapulid worms, while later the onychophorans, and a few other minor phyla were added.

These two superphyla, Lophotrochozoa and Ecdysozoa, have since been supported by several detailed molecular analyses, and together they form the somewhat modified traditional group Protostomia (Bourlat et al. 2008; Dunn et al. 2008; Helmkampf et al. 2008; Telford et al. 2008; Rota-Stabelli et al. 2010). Unlike this radical modification of the phylogenetics of Protostomia, their traditional sister-group amongst the Metazoa, Deuterostomia, has remained a monophyletic taxon that is well supported by molecular evidence for hemichordates, echinoderms, and chordates.

As reviewed recently by Dohrmann and Wörheide (2013), at the basal level of invertebrates, molecular evidence has not yet resolved unambiguously the interrelationships of Bilateria to the non-bilaterian metazoans, Cnidaria, Ctenozoa, Porifera, and the very simple, small species *Trichoplax adhaerens* that constitutes the phylum Placozoa. In most studies the sponges, Porifera, consistently show up as the sister-group of the other three plus Bilateria, but the recent publication of the complete genome of the ctenozoan *Mnemiopsis* indicates that it is actually this group that is the sister-group of the rest of the metazoans (Ryan et al. 2013).

Aside from this basal level, the phylogenetic interrelationships of the eumetazoan superphyla can certainly be regarded as sufficiently well-established as to provide the framework within which the fossil evidence should be interpreted (Bourlat et al. 2008; Edgecombe et al. 2011).

7.2 The Cambrian explosion

Any account of the origin of the invertebrate phyla must at some point face up to the extraordinary event that occurred between about 540 and 520 Ma and is referred to as the Cambrian explosion (Erwin et al. 2011; Erwin and Valentine 2013). In *On the Origin of Species*, Darwin himself famously remarked uncomfortably upon the apparently sudden appearance in the fossil record of many major animal taxa in what Adam Sedgwick had recently labelled the Cambrian period. The mystery has if anything deepened since then: more and more superbly preserved Cambrian fossils continue to attest to the geologically speaking abrupt appearance in the fossil record of this hugely disparate radiation, whilst molecular-based estimates consistently point to substantially pre-Cambrian divergence dates between the modern phyla. Arguments continue about the extent to which the Cambrian explosion reflects an extraordinarily rapid phase of major diversification, and the extent to which there was a very long prior episode of divergent evolution has left virtually no fossilized trace: most would now accept that the answer is a combination of both. Either way, there are implications for the mechanism of major evolutionary change. In addition to the morphology of the fossils, addressing the conundrum of the Cambrian explosion involves evidence about the late Phanerozoic and early Cambrian environments and about the genetics of body plan development. The latter two aspects are discussed elsewhere; for the moment, the present concern is the extent to which Cambrian fossils illustrate the evolution of crown phylum characters.

Multicelled organisms of a variety of eukaryote taxa have existed for at least half the Proterozoic, with possible records from about 2 billion years ago. However, virtually without exception these were simple colonial or filamentous forms and the only possible metazoans were minute, spicule-bearing sponges (Butterfield 2009). There is one form that has been claimed as a very early bilaterian metazoan. *Vernanimalcula* is from the Doushantuo Formation of China, which dates at about 600 Ma, although it is possibly younger. It is described as very small, about 0.2 mm, and possessing a series of organs characteristic of bilaterian animals (Fig. 7.3a). These include a paired coelom, mouth, muscular pharynx, digestive tract, anus, and possible sensory pits on the surface (Chen et al. 2004; Petryshyn et al. 2013). If correctly interpreted, *Vernanimalcula* is a very significant fossil for the light it throws on the date and anatomy of basal bilaterians. However, there is a good deal of doubt about the interpretation of the specimens and most of the alleged structures can be equally well explained as artefacts of entirely geochemical mineralization processes (Bengtson et al. 2012).

Apart from this possibility, the earliest undoubted metazoan fossils occur in the Ediacaran biota (Vickers-Rich and Komarower 2007; Xiao and Laflamme 2009). This is named after the original locality in Australia but is now known from some 40 localities worldwide, with ages ranging from about 560 Ma to the start of the Cambrian at 542 Ma. The diversity of the biota is quite modest, with around 100 species, but its disparity is astonishing (Fig. 7.3b). Most specimens are preserved as impressions of organisms, and traces of trackways and burrows in sandstones and other medium-grained clastics, although a few body fossils have been described. They are extremely difficult and controversial to interpret, and there continue to be numerous arguments about whether particular species are stem members of modern phyla such as cnidarians, annelids, and molluscs, or at the other extreme whether they are unrelated to anything known since. Candidates for stem-group status include the fairly convincing mollusc *Kimberella* (see following, section 7.3), and the very unconvincing possible annelids *Dickinsonia* and *Spriggina*, in which the impressions of body annulations have been taken as segmental boundaries and parapodia. Others, such as the fronded rangeomorphs *Rangea* and *Charnia* are built to a body plan

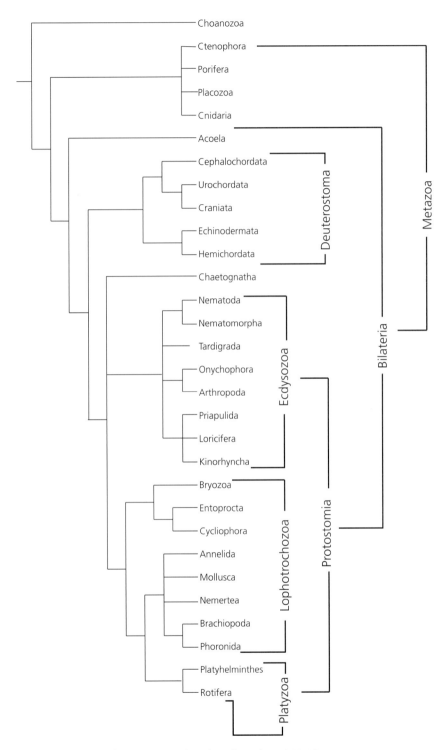

Figure 7.2 Molecular-based phylogeny of the invertebrates. (Based on Edgecombe et al. 2011.)

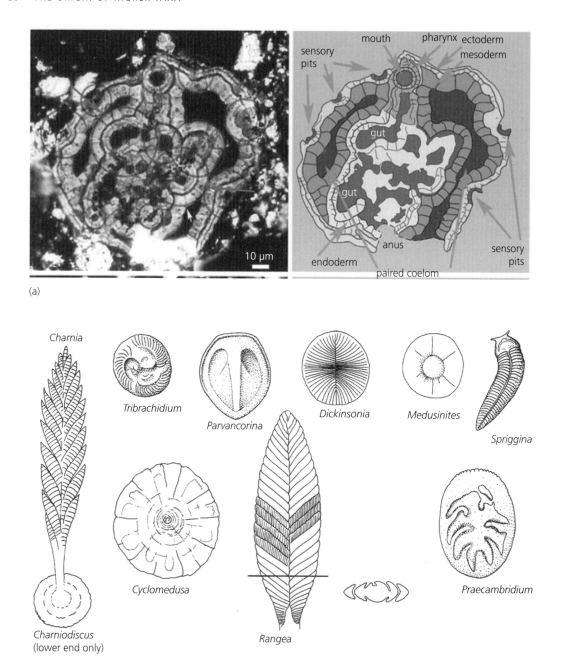

Figure 7.3 (a) Section and interepretation of *Vernanimalcula*, a possible though doubtful Precambrian diploblast. (b) A selection of ediacaran taxa. ((a) from Bengtson et al. 2012; (b) from Clarkson 1979.)

unlike any other metazoans, as is the irregularly bag-like *Ernettia*, and the one-time proposed arthropod *Parvancorina*. All lack clear synapomorphies of crown groups. The discoidal fossils, once attributed to the Cnidaria, are now generally accepted to be an assortment of detached holdfasts of larger organisms and probably colonial bacteria and fungi.

Seilacher's (2007) interpretation is that not only are most of the ediacaran taxa not members of recognized crown phyla, but at least many of them are not even members of a monophyletic Metazoa. He named the group Vendobionta and proposed that they were built on an entirely different plan to metazoans, involving a quilted structure. The surface of the organism was flexible and could expand during growth, and the body itself consisted of small bag-like compartments that were added to during growth. He suggested that they may also have been coenocytic rather than conventionally cellular in structure. Nutrition was by surface absorbtion. Retallack (1994) was responsible for an even more egregious suggestion, namely that the vendobionts were not animals at all but lichens, a theme he recently returned to along with arguments that at least some of the biota was terrestrial rather than aquatic (Retallack 2013, but see Xiao 2013).

Fascinating as the Ediacaran biota is in its own right, neither the bizarre nature of many of its members, nor the lack of clear phylogenetic relationships amongst them adds much to understanding how they evolved as new higher taxa, and they disappeared without trace close to the start of the Cambrian period. The cause of the extinction of the Ediacaran biota is difficult to assess, not least due to the lack of exact dating of the different localities in which it occurs worldwide. The apparent abruptness of its disappearance suggest that it was the result of a mass extinction event, but some have suggested that it was a more gradual decline caused by competition and predation by the new taxa that evolved at this time (Laflamme et al. 2012). Whatever the case may be, the start of the Cambrian is marked by the first appearance of mineralized skeletal remains of organisms (Bengtson 2004; Maloof et al. 2010). By the first part of the Cambrian, the Tommatian, there is a rich disparity of skeletal remains referred to collective as the 'small shelly fauna' (Fig. 7.4). In most cases the specimens themselves are not very informative about the morphology and phylogenetic relationships of the organisms that possessed them. There are very small halkeriid, molluscan and brachiopod-like shells, disarticulated echinoderm ossicles and arthropod-like skeletal elements, plus a variety of sponge spicules, pieces of test, tooth-like remains, and tubular structures.

The small shelly fauna bears witness to the extent of the biodiversity at the start of the Cambrian but the specimens in themselves reveal little about the relationships, origin, and evolution of early Cambrian organisms. However, thanks to the spectacular Cambrian Lagerstätten containing exquisitely preserved soft-bodied as well as hard-skeletonized organisms, dominated by arthropods and sponges, a great deal is now known about the nature of the Cambrian animals, and about the environmental conditions of the period. The most important Lower Cambrian Lagerstätten are the Chengjiang and others in South China (Zhang, X.-L., et al. 2008), the rather less rich Sirius Passet in Greenland which contains about 40 species (Peel 2010), and the still relatively little studied Emu Bay Shale in Australia (Gehling et al. 2011). These are followed in time by the classic Burgess Shale in British Columbia which is Middle Cambrian, and the Upper Cambrian Orsten fauna, first described from Sweden but now known to have been virtually global in distribution. Between them, they include the earliest well-preserved members attributed to all the important invertebrate phyla.

Out of the considerable number of early invertebrate fossils of the Lower and Middle Cambrian, there are several that have been and continue to be put forward as candidates for stem membership of crown phyla, particularly Mollusca, Arthropoda, Echinodermata, and Chordata. However, the fact is that even amongst the relatively undisputed stem members, few are particularly helpful for deciphering the pattern and sequence of acquisition of the characters of the crown phyla. They tend to possess either very few of the crown taxon characters which are furthermore confusingly overlain by extreme apomorphies of their own, or else they possess so many of them that they are scarcely distinguishable from crown membership themselves. The fossil record is therefore relatively unhelpful for throwing light on how the basic body plans of modern phyla were assembled. This is why the relatively little that actually is understood about the origin of the characteristic morphologies of almost all animal phyla is mainly inferred from comparative anatomical, molecular, and developmental studies. The best case is certainly the Arthropoda (Edgecombe and Legg 2014), not only because it is the best represented phylum in the Early and Middle Cambrian, but also because many of the fine morphological features preserve particularly well thanks to the chitinous skeleton of the segments and appendages. Fossils do contribute significant information

Figure 7.4 The Early Cambrian 'small shelly fauna'. (From Bengtson 2004.)

about the timing of early invertebrate evolution and something of the conditions on earth at the time, although even here conclusions are not infrequently vague and contradictory.

In addition to the phyla, the individual classes within them generally differ sufficiently markedly from one another to count as distinct higher taxa in their own right, each with its own body organization as a variant on that of the phylum in general. Again, as with the phylum level, there are actually very few stem members of the constituent classes that differ sufficiently from both the ancestral and the respective crown groups to be able to shed much light on the way these classes themselves evolved.

7.3 Lophotrochozoa: Mollusca

There are seven classes of living molluscs: the worm-like, unshelled Aplacophora, the multi-plated chitons or Polyplacophora, the Bivalvia, the single-shelled Monoplacophora, the Gastropoda, the Scaphopoda, and the Cephalopoda. The monophyly of the living molluscs is scarcely doubted, despite the considerable disparity amongst the classes, on the basis of both morphological and molecular evidence. The only recent doubt has arisen from Wilson and colleagues' (Wilson et al. 2010) multigene analysis using several analytical methods, in which the Aplacophora was found to be paraphyletic: one of its subgroups, the Solenogastres came out as the sister-group of the sipunculids, nested within the annelids, while the other, Caudofoveata, remained within Mollusca, although unexpectedly as the sister-group of the cephalopods (Fig. 7.5a). However, other recent multigene analyses (Kocot et al. 2011; Vinther et al. 2012; Vinther 2015) do support monophyly.

The interrelationships amongst the molluscan classes are still disputed, and the disagreements bear directly on the inferred nature of the ancestral mollusc and what its characters were. Notably this relates to the position of aplacophorans and the consequent significance of their shell-less, vermiform nature. Some authors have concluded that the aplacophorans lie at the base of the crown mollusc tree, either as a monophyletic or a paraphyletic taxon, which implies that the molluscan ancestor lacked a muscular foot and calcified shell, but instead progressed in a worm-like fashion (e.g. Haszprunar 2000). Others propose what is termed the 'aculiferan' hypothesis, in which the aplacophorans are related to the serially shelled polyplacophorans, and together they constitute the sister-group of the remainder of the molluscs, referred to as the Conchifera (e.g. Sigwart and Sutton 2007). The obvious inference in this case is that the lack of a shell and a foot in the aplacophorans is due to secondary loss, and that the molluscan ancestor did possess these characteristics. The molecular studies quoted earlier are contradictory on this issue. Wilson and colleagues' (2010) conclusion (Fig. 7.5a) supports the worm-like ancestor, while Kocat (2011) and Vinther and their colleagues (2012) support the Aculifera hypothesis, but with the proviso in the latter that the shelled classes are not a monophyletic group because of the relationship of the cephalopods to the aculiferans (Fig. 7.5b and c).

Another fundamental issue concerns the relationships of the Monoplacophora. These were once considered the most primitive living molluscs because of the serial repetitions of some of their internal structures, namely the pedal muscles, ctenidia, excretory organs, and nerves, which was assumed to

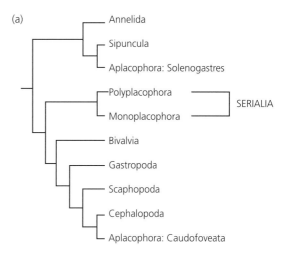

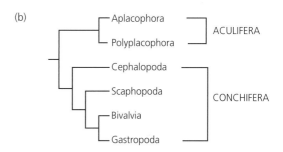

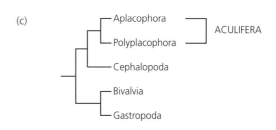

Figure 7.5 Three competing molecular phylogenies of molluscs. (a) The Serialia theory of Wilson; (b) the aculiferan-conchiferan theory of Kocat et al. 2011; (c) the version of the aculiferan theory of Vinther et al. (2012).

be a primitive retention of a degree of metameric segmentation. Also they appeared very early, in the Early Cambrian small shelly fauna. However, the various kinds of simple, cap-shaped shells found at this time are not at all well understood (Peel 1991), and in any case it is virtually certain on developmental grounds that the arrangement of the organs in the living monoplacophorans does not represent true metamerism at all, and that it is just as likely to be autapomorphic for the class. The similarity of many morphological features of monoplacophorans to those found in the rest of the mono- and bivalved classes has led most to accept a monophyletic clade, Conchifera, to include Monoplacophora with Gastropoda, Scaphopoda, Bivalvia, and Cephalopoda. In contrast, Wilson and colleagues (2010) found some, though weakish support for a relationship between Monoplacophora and Polyplacophora (Fig. 7.5a), and the two living groups both exhibit serial repetition of certain parts. If corroborated, this would suggest that serial repetition is a derived condition of this clade, which they named Serialia, leaving the ancestral mollusc supposedly non-serial.

In view of the relatively poor resolution of the interrelationships of the molluscan classes, it might well be hoped that the fossil record of what is one of the most fossilizable phyla of all would throw helpful light on the polarities and sequence of acquisition of molluscan characters leading to the body plan of the crown group and its included classes. What is actually found is an intriguing but on the whole frustratingly small number of Cambrian and indeed Precambrian candidates for stem mollusc status, over which much disagreement prevails. The problem is essentially that of distinguishing true homology from superficial similarity of soft structures preserved in little more than external, distorted outline, or as ambiguous indications on harder parts of where soft tissues once attached. The subject is not without a degree of prejudice and imagination.

7.3.1 *Kimberella*

Originally described from the Late Precambrian Ediacaria fauna by Glaessner and Wade (1966, Glaessner 1984) as a scyphozoan, *Kimberella* (Fig. 7.6a) was important from the start in establishing the belief that certain modern phyla were present at this very early time. However, new and much better preserved material of a similar date has since been discovered in northern Russia and this allowed a more detailed description by Fedonkin and Waggoner (1997; Fedonkin et al. 2007). In size, *Kimberella* varies from 3 mm to 150 mm in length and appears to have been a benthic, bilateral organism. These authors interpreted it as a very basal mollusc on the basis of the presence of a single-valved, though non-mineralized dorsal shell, a broad, creeping foot, and signs of serially arranged musculature running ventrally to the foot. They also claim evidence for a buccal mass, but no radula is seen in any of the specimens. However, they do suggest that certain scratch marks on the substrate were made by a molluscan-like radula, which they presumed must have been possessed by this animal. Apparent crenulations around the margin of the animal may conceivably be serially arranged simple respiratory ctenidia. On their interpretation, *Kimberella* is indeed the basalmost stem mollusc known. It implies that the earliest of the mollusc characters to evolve included a creeping foot, dorsal protection by at least a toughened cuticle though not at this stage a mineralized shell, and serial repetition of musculature and possibly respiratory organs. These features in turn suggest that metameric segmentation was not part of molluscan history, and that the worm-like locomotion of the aplacophorans is secondary.

Since, like the Ediacaran biota generally, *Kimberella* is preserved only as impressions in sandstone, the anatomical resolution is poor, and the structure remains open to alternative interpretations. Butterfield (2006) expressed great scepticism and went no further than to recognize it as a 'probable bilaterian'. Conway Morris and Caron (2007) felt that the proposed homology between the non-mineralized dorsal surface with the spicules or shell of molluscs was doubtful, and that *Kimberella* could as well be a stem lophotrochozoan as a specifically stem mollusc. Ivantsov (2009) studied the preservation of a large collection of new Russian specimens and showed that the body was capable of extensive elongation, up to twice the resting length, and therefore could not have been covered by a single mollusc-like dorsal shell, but may have borne small, separate sclerites. He accepted that it may well be a stem mollusc, but his interpretation suggests a rather different ancestral state for molluscs, more akin to the worm-like aplacophoran than to the single-shelled conchiferan.

7.3.2 *Odontogriphus*

Amongst the spectacular Middle Cambrian Burgess Shale Lagerstätten fauna, *Odontogriphus* (Fig. 7.6b) has variously been associated with brachiopods, phoronids, ectoprocts, and chordates in the past. More recently Caron and colleagues (2006) identified it as a stem mollusc on the basis of a large number of new specimens. In particular, they recognize a molluscan radula formed of two rows of denticulated teeth surrounded by what appears to be a circular mouth. The ventral surface of the animal suggests a muscular foot, and it is surrounded by serially repeated structures lying in a shallow groove. These are interpreted as ctenidia contained within a mantle cavity. The dorsal surface is not mineralized, but judging by its state of preservation was a tough cuticle. *Odontogriphus* is therefore comparable in its general form with the molluscan interpretation of *Kimberella*, but with the additional molluscan characters of a mantle groove housing repeated ctenidia.

The principle objections to *Odontogriphus* as a stem mollusc were raised by Butterfield (2006; see Caron et al. 2007 for their vigorous response), who doubted whether the structure of the supposed radula is truly molluscan. It differs from an Early Cambrian radula that he himself later described (Butterfield 2008) and which he regarded as indisputably molluscan, and is equally comparable to the mouthparts of certain polychaete annelids. He was also sceptical of the presumed homology between the lateral gill-like structures and molluscan ctenidia. Thus Butterfield's conclusion was that *Odontognathus* could equally well be a stem lophotrochozoan or related to any of the lophotrochozoan classes such as the annelids. Since this debate, Smith (2012) has made a very detailed study of the dentition of *Odontogriphus* and of *Wiwaxia* (see section following), showing clearly that it is molluscan in form. The individual teeth are set in similar-sized transverse rows in a membrane, and are added to from behind. More precisely, the nature of the radula corresponds to those of deposit-feeding molluscs such as chitons.

7.3.3 *Halkieria*, *Wiwaxia*, and *Orthrozanclus*: Halwaxiidae

The Early Cambrian *Halkieria* (Fig. 7.6c) from Sirius Passet in Greenland (Conway Morris and Peel 1990, 1995) is an elongated animal with a soft foot. The dorsal surface is covered with a mass of small, sharp, calcareous sclerites and the foot is surrounded by a further row of slightly elongated ones. In addition to these, there is, curiously, a single calcareous shell at each end of the body. There is no preserved radula or any other mouth parts, and neither mantle groove nor ctenidia are found. From the start it was proposed that *Halkieria* was related to the lophotrochozoan group, and in particular to the brachiopods and annelids (Conway Morris and Peel 1995). Indeed, Conway Morris (1998) later went so far as to suggest that the brachiopods evolved from a primitive halkieriid by bending the body so that the two shells became opposed to one another, thereby creating a bivalved condition. If this is the case, *Halkieria* is a stem brachiopod. Others however, notably Vintner and Nielsen (2005), were convinced of the molluscan nature of the shell and sclerites, and later (Vinther 2009) it was claimed that the similarities are specifically with aplacophorans and polyplacophorans which implies that *Halkieria* is a crown-group mollusc and a member of the proposed taxon Aculifera mentioned earlier.

Wiwaxia (Fig. 7.6d) is one of the classic forms from the Middle Cambrian Burgess Shale. Like *Halkieria*, the dorsal surface is covered by short sclerites, but there is a row of much longer ones down the midline and it lacks the two compact shells of the earlier genus. From its very first description in the early twentieth century (Walcott 1911), *Wiwaxia* was implicated in the origin of annelids, on the basis of similarity of its sclerites to polychaete setae, and as recently as 1990 Butterfield (1990) explicitly placed it in the class Polychaeta. However, what Walcott took to be the homologue of setae, and Butterfield took to be annelid-like feeding denticles, others have interpreted respectively as molluscan sclerites and a radula, and therefore that it is with the Mollusca that its affinities lie (Conway Morris 1985; Eibye-Jacobsen 2004; Caron et al. 2006). Furthermore, *Wiwaxia* shows no evidence of the metemeric segmentation that might be expected in a stem annelid.

It is clear that *Wiwaxia* and *Halkieria* are closely related, and are included in a taxon Halwaxiidae. *Orthrozanclus* (Fig. 7.6e) is another genus from the Burgess Shale (Conway Morris and Caron 2007). It has characters of both of the other two genera,

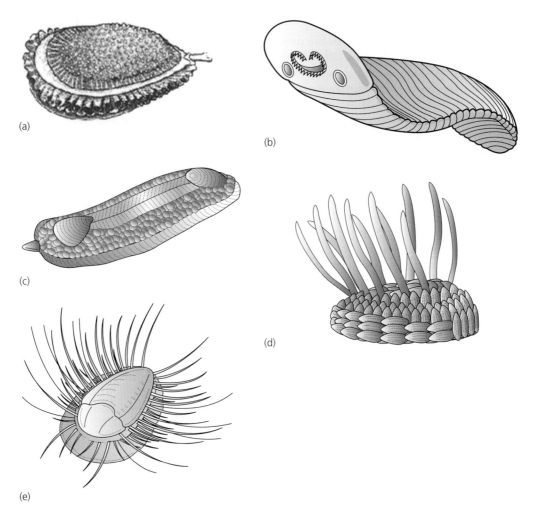

Figure 7.6 Possible stem molluscs. (a) *Kimberella* (From Fedonkin 2007); (b) *Odontogriphus*; (c) *Halkieria* (from Conway Morris and Peel 1995, reproduced with permission from the Royal Society); (d) *Wiwaxia* (modified from Conway Morris 1985); (e) *Orthrozanclus* (from Conway Morris and Caron 2007, reproduced with permission from the American Association for the Advancement of Science).

notably an anterior although not a posterior shell, and elongated sclerites like those of *Wiwaxia*. How exactly the halwixiids relate to the lophotrochozoan phyla remains debatable. Caron and colleagues (2006; Caron et al. 2007) propose, and Conway Morris and Caron (2007) accept as one possibility that they are stem molluscs, a view supported by the presence of the radula and supported by Smith's (2013a) recent study of the sclerites and other internal structure. If correct, this view demonstrates that the ancestral mollusc possessed a dorsal sclerotome of individual sclerites rather than a continous

shell. However, Conway Morris and Caron (2007) offer an alternative arrangement, namely that halwixiids constitute the stem group of annelids plus brachiopods, which would imply that the radula was not actually a molluscan character, but a plesiomorphic one that had been lost in the annelid–brachiopod lineage It also contradicts certain molecular studies pointing to a sister-grouping of brachiopods and molluscs to the exclusion of annelids (Helmkampf et al. 2008; Wilson et al. 2010). At the other extreme is Vinther's (2009) view that wiwaxiids are crown molluscs related to the aculiferan classes Aplacophora and Polyplacophora. Yet another complication is the demonstration that the aragonitic microstructure of the sclerites of *Halkieria* is virtually identical to that found in a group of simple, sponge-like Cambrian organisms called chancellorids (Porter 2008). It may be that chancellorids are basal bilaterians and the presence of sclerites was a character of Bilateria as proposed by Bengtson (2004), or that chancellorids are highly derived, simplified halkieriids, or of course that the similarity is merely convergent in chancellorids.

Assuming, and the caveats mentioned in the last paragraph indicate it is a bold assumption, that the crownward sequence of branching off the molluscan stem lineage was *Kimberella, Odontogriphus,* Halwaxiidae, Crown Mollusca) then a tentative sequence of acquisition of some of the crown mollusc characters from a hypothetical lophotrochozoan ancestor can be inferred:

(i) Creeping, muscular foot with serially arranged pedal musculature; simple pre-ctenidial respiratory extensions in a ventral groove around the body; muscular buccal mass; possibly a simple radula.
(ii) Definite radula; distinct ctenidia in a pallial (mantle) groove.
(iii) Calcareous spicular sclerites embedded in the dorsal surface, and at least a partial true shell.
(iv) Single, continuous calcareous shell over the body mass.

Taken at simple face value the sequence supports the Aculifera hypothesis that the ancestral crown mollusc was a compact, non-segmented but partially serially organized organism, with a muscular foot rather than a vermiform, aplacophoran-like form. However, in the face of the doubts expressed about the homologies of the preserved structures and taxonomic positions of these remote fossil forms, plus the absence of any information at all about many of their characters, little of general value concerning the pattern of acquisition of the derived characters of the Mollusca emerges from their study.

7.3.4 Crown Mollusca

Notwithstanding the molecular evidence (Fig. 7.5) that the Aplacophora, with or without association

with Polyplacophora diverged close to the base of the molluscan radiation, they appear relatively late in the fossil record, in the early Ordovician. Polyplacophora too is a relatively late arrival, since they are not known until the Late Cambrian onwards (Runnegar et al. 1979; Vendrasco and Runnegar 2004). However, recent fossil evidence does throw some light on the evolution of these two classes. Sutton and colleagues (2004) described a Silurian worm-like form *Acaenoplax* (Fig. 7.7) whose soft parts and three-dimensional

Figure 7.7 *Acaenoplax.* (Redrawn from Sutton et al. 2004 and reproduced with permission from the Palaeontological Association and John Wiley & Sons.)

structure are preserved. It combines characters of both aplacophorans and polyplacophorans. Like aplacophorans, the overall shape is vermiform rather than flattened with a muscular foot, and it appears to possess an anterior mouth and posteriorly displaced mantle cavity. On the other hand, along the dorsal surface there is a series of somewhat polyplacophoran like aragonitic plates, and around the sides lie spicules, also aragonitic, that resemble those of both polyplacophorans and aplacophorans. It also possesses a number of unique characters of its own, notably the rather annelid-like median groove along the underside. Further evidence came from a second Silurian form, *Kulindroplax* (Sutton et al. 2012), which is similar except that the serially arranged shell valves are much larger and therefore more chiton-like. Cladistic analysis indicates that these fossils are indeed aplacophorans, and supports the relationship between Aplacophora and Polyplacophora as the clade Aculifera at the base of the crown mollusc tree, as concluded from some molecular studies (Vinther et al. 2012). Furthermore, this arrangement supports the notion that the ancestral molluscan condition is shelled and muscular-footed, rather than worm-like: the latter is inferred to be a secondarily derived condition in the Aplacophora.

Molecular studies of the relationships of cephalopods are incongruent at present, with some evidence pointing to a sister-group relationship with the clade Aculifera, the aplacopherans plus polyplacophorans (Vinther et al. 2012) and other evidence favouring a sister-group relationship with a clade Conchifera, the gastropods, bivalves, and scaphopods (Kocot et al. 2011). Fossil evidence bearing on the origin of cephalopods has until recently mostly been limited to shell comparisons, which can indicate very little about most of the character transformations. A number of authors have argued for a third phylogeny, a monoplacophoran relationship (Kröger 2007; Kröger et al. 2011) on the grounds that certain early Cambrian monoplacophorans have a curved shell and a serial arrangement of shell muscles resembling those of the earliest undisputed members of the Cephalopoda (Fig. 7.8c–f), which are the Late Cambrian ellesmeroceridan nautiloids (Fig. 7.8d). Dzik (2010) disputed this claim and proposed a different early mollusc of unknown affinities called *Turcotheca*

(Fig. 7.8b) as ancestral to cephalopods. It consists of a high, slightly curved and compressed shell comparable to that of ellesmeroceratidans.

Recently and potentially a good deal more informatively, however, Smith and Caron (2010) argued that the Middle Cambrian Burgess Shale form *Nectocaris* (Fig. 7.8a) is a stem cephalopod. According to their interpretation of its structure, it possesses two cephalopod-like tentacles and an axial mantle cavity containing a series of paired gills and opening anteriorly as a jet-propelling funnel behind the mouth. Other putative cephalopod features are a pair of stalked, camera rather than compound eyes, and lateral fins, but cephalopod characters not present include a horny beak or radula, eight tentacles, and a calcified shell. If their interpretation is correct, the nectocaridids would indeed be a most informative stem cephalopod. Taken at face value they imply, for example, that controlled jet propulsion preceded the evolution of a buoyancy shell. However, considerable disagreement with this interpretation of *Nectocaris* was immediately expressed. The evidence for a cephalopod-like mantle cavity and funnel was found unconvincing and better seen as part of the gut, perhaps with an anterior proboscis-like structure at the front (Kröger et al. 2011; Mazurek and Zatoń 2011; Runnegar 2011). Other objections were to the assumption that the shell must have already been secondarily lost, assuming a conchiferan origin of cephalopods as all other evidence points towards. Furthermore, the fins and possibly the eyes are uniquely coleoid rather than basal cephalopod characters and would not therefore be expected in a stem cephalopod. Smith (2013) returned to the argument and on the basis of his new investigations identified what he believed to be mouthparts resembling those of the supposed stem molluscs *Odontogriphus* and *Wiwaxia*, and demonstrated a specifically cephalopod-like structure of the phosphatized gill remnants in the supposed atrial cavity. He also argued that the form of the funnel is functionally suitable for a jet-propelled organism of the small dimensions of *Nectocaris*, and accepts that the eyes and fins may be convergent on those of the coleoid cephalopods that appeared much later, in the Devonian.

As much as any other taxon, *Nectocaris* epitomizes the difficulty of interpreting the phylogenetic position of Cambrian metazoans even where traces

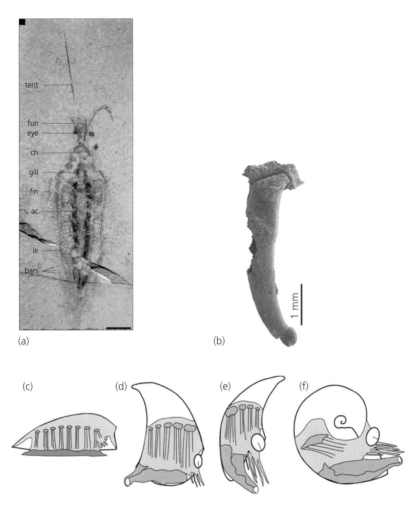

Figure 7.8 Possible stem cephalopods, (a) *Nectocaris* as preserved (from Smith and Caron 2010, reproduced with permission from Nature Publishing Group); (b) *Turcotheca* (from Dzik 2010). (c–f): a possible transformation series from (c) a monoplacophoran, via (d) an ellesmeroceridan, and (e) an oncoceridan, to (f) *Nautilus* (from Kroger 2007 reproduced with permission from the Palaeontological Association and John Wiley & Sons).

of soft tissues are preserved. Currently, different respective proponents argue for the view that it is a stem cephalopod (Smith and Caron 2010; Smith 2013), an independent lineage of lophotrochozoan (Kröger et al. 2011), 'a stem group member of some other animal phylum' (Runnegar 2011), and even possibly related to the anomalocaridids, a presumed stem euarthropod group (Mazurek and Zatoń 2011).

The earliest gastropods, members of what is now the most diverse mollusc class, occur in Lower Tommotian rocks of the Early Cambrian. Little can be inferred about their soft structures and indeed

they can be difficult to distinguish from contemporary monoplacophorans (Parkhaev 2007), and certainly nothing at all is known about the origin of torsion, perhaps the single most definitive character of crown gastropods.

Possible fossil scaphopods occur in the Ordovician, although the picture is confused by a tendency of several mollusc taxa to evolve a tubular, open-ended shell in association with infaunal life (Peel 2006). At any event, no specimens have been described that throw any light on the evolution of the scaphopodan version of the molluscan body plan.

The small bivalves *Fordilla* and *Pojataia* are well known from Early Cambrian 'small shelly beds' (Runnegar and Pojeta 1972; Fang 2006; Elicki and Gursu 2009). What little can be made out of their structure indicates a significantly more primitive organization than crown bivalves. The bivalved shell has a simple hinge and ligament and the muscle scars on the inner surface of the valves indicates that the adductor musculature is relatively poorly differentiated, and the foot probably emerged ventrally, rather than anteriorly as in later, crown bivalves.

7.4 Ecdysozoa: Arthropoda

Despite the wide disparity of segmental and tagmatization patterns and appendage morphology amongst arthropod classes that led some earlier influential systematists to conclude they are polyphyletic (Manton and Anderson 1979), molecular evidence overwhelmingly supports their monophyly (Bourlat et al. 2008; Budd and Telford 2009; Rota-Stabelli et al. 2010). The crown group which consists of the hexapods, crustaceans, myriapods, and chelicerates, plus the extinct class Trilobita, is generally referred to as the Euarthropoda although its relationship to other members of the Ecdysozoa are not confidentally determined (Fig. 7.9). Onychophora, the velvet worms, is the best supported candidate for the living sister-group of euarthropods, and is frequently included with them as the Panarthropoda. However, another contender is Tardigrada, the water bears (Budd and Telford 2009; Rota-Stabelli et al. 2010; Smith and Ortega-Hernandez 2014), while the priapulid worms are

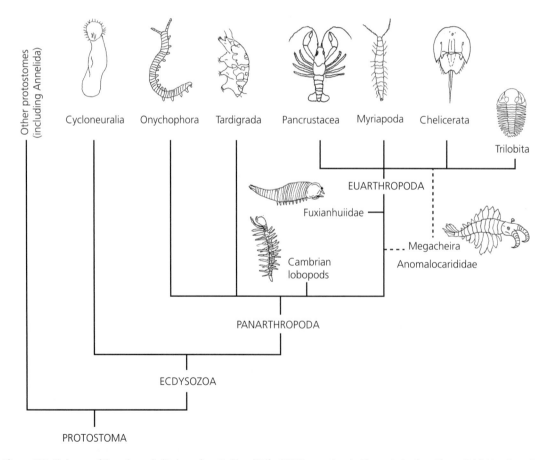

Figure 7.9 Phylogeny of the arthropods. (Redrawn from Budd and Telford 2009, reproduced with permission from Nature Publishing Group.)

also considered to be close. Bourlet (2008) and Rota-Stabelli and their colleagues (2010), for example, find Priapulida to be the sister-group of a panarthropoda that includes both onychophorans and tardigrades.

Notwithstanding the present uncertainty about the detailed interrelationships within the ecdysozoans, the establishment of the monophyly of the modern arthropod classes allows some of the morphological characters of their hypothetical common ancestor to be inferred. In addition to the relatively few morphological characters that define the Ecdysozoa as a whole, mainly moulting of the cuticle, a trilaminate cuticle structure, and a terminal mouth (Budd and Telford 2009), Panarthropoda probably possessed an organized level of segmentation, with each segment bearing a simple pair of appendages. In this light, there are a number of possible stem-group arthropods in the fossil record that show this basic structure but lack the synapomorphies of any of the crown arthropod classes, as reviewed most recently by Edgecombe and Legg (Edgecombe and Legg 2014). Many occur as soft-bodied fossils in Cambrian Lagerstätten, and all are the subjects of alternative interpretations of both structure and phylogenetic position. The most basal are rather loosely described as lobopodiforms. They formed an abundant and diverse component of the Lower and Middle Cambrian faunas, forming a paraphyletic group that presumably includes the ancestor of the panarthropods and probably of the euarthropods (Liu and Dunlop 2014). The structure and detailed homologies of the different taxa are difficult to interpret and their interrelationships poorly resolved, notwithstanding a number of attempts (Daley et al. 2009; Liu et al. 2011). Several are claimed to have a specifically stem arthropod relationship, such as the Chengjiang genus *Diania* (Fig. 7.10a). Liu and colleagues (Liu, Steiner et al. 2011) described it as combining soft, non-sclerotized body segments with hard-jointed limbs, and therefore a possible stage in the evolution of arthropods. Another example, from the Sirius Passet fauna, is *Kerygmachela* (Fig. 7.10b), which Budd (1999) described as possessing a well-defined head region with long frontal appendages, followed by eleven segments each consisting of a pair of lobe-like limbs surmounted by paired lateral lobes that

he interpreted as gills. He suggested that this limb structure represents an incipient version of the biramous arthropod limb, a condition seen to a more developed extent in *Opabinia*. However, given the diversity of the Cambrian lobopodans coupled with the difficulty of deciphering their structure, no such attributions can be confidently upheld.

7.4.1 *Opabinia*

No fossil has played a more central role in trying to unravel the origin of arthropods than the Burgess Shale *Opabinia* (Fig. 7.10c). It is a segmented organism bearing a peculiar median flexible appendage at the front and a tail structure or telson formed from the last three segments. Budd (1996; Budd and Daley 2011) interpreted it as a stem euarthropod on the grounds that it possesses several euarthropod characters, as did Legg and colleagues (2013). There is a set of five stalked eyes that are judged from their size to have been compound. Behind the head, each segment has a pair of ventral walking lobopods as in lobopodans generally, each terminating in a tiny claw. The lobopod is surmounted by a dorsal, leaf-like lateral lobe regarded as a gill, and the combination of the lobopod and gill is interpreted as an initial evolutionary stage of what was to become a euarthropol biramous limb. However, interpretation of the structure is difficult and others disagree that there are walking limbs as well as lateral lobes, the latter presumably for swimming with gill elements attached to the dorsal surface (Zhang and Briggs 2007, see Budd and Daley 2011). At the anterior end of the animal, the median appendage is not as such found in any euarthropods, but the pair of tiny terminal claws it bears are supposedly a further step in the process of arthropodisation.

7.4.2 Anomalocaridida (Radiodonta)

The next crownward clade of euarthropods is believed by many to be the Anomalocaridida (Fig. 7.11a and b), the 'great appendage' arthropods, which are found throughout the Cambrian, including Chengjiang, Sirius Passet, and Burgess Shale (Daley et al. 2009). They are evidently related to *Opabinia*, because they share the segmental appearance and presumed compound eyes, tail fan, and presence, if correctly

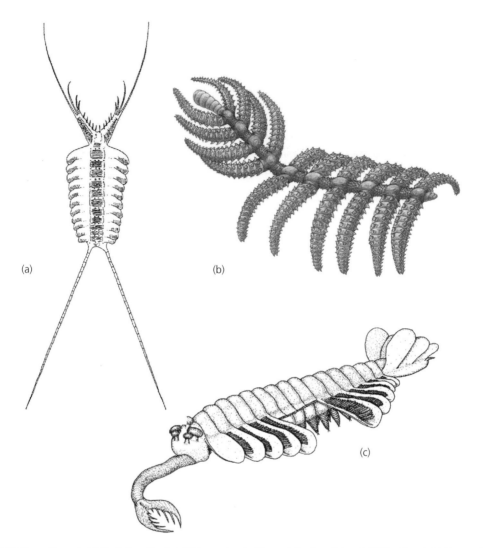

Figure 7.10 Stem arthropods. (a) *Diania* (from Liu et al. 2011 reproduced with permission from Nature Publishing Group); (b) *Kerygmachela* (from Budd 1993 reproduced with permission from Nature Publishing Group); (c) *Opabinia* (from Budd 1996 reproduced with permission from John Wiley & Sons).

identified in *Opabinia*, of serial appendages with both lateral lobe and ventral lobopod. However, anomalocaridids have some additional euarthropod characters, notably the fusion of these two parts of the appendages to form a single, biramous trunk limb, and its sclerotization. The 'great' or frontal appendages themselves consist of a pair of elongated, jointed structures lying alongside the mouth. The articulated structure extends right down to a basal peduncle, and the strong sclerotization leaves

little doubt that they are essentially arthopodan in nature. The size of the great appendages varies from species to species, presumably reflecting differences in diet and even including possible filter-feeding (Vinther et al. 2014). Those of the giant *Anomalocaris* are about half the body length; others such as *Hurdia* (Fig. 7.11b) less than half this (Daley et al. 2009). The mouth is a peculiar structure unlike any euarthropod, being circular and surrounded by 32 tooth plates and numerous tiny teeth.

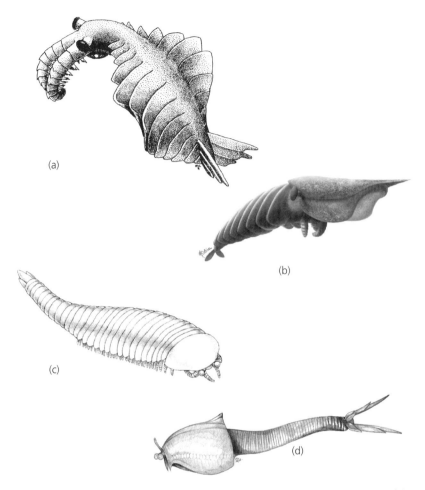

Figure 7.11 (a) *Anomalocaris* (from Collins 1996); (b) *Hurdia* (from Daley et al. 2009 reproduced with the permission of the American Association for the Advancement of Science); (c) *Fuxianhuia* (from Hou et al. 2007); (d) *Nereocaris* (Legg et al. 2012 reproduced with permission from the Royal Society).

7.4.3 *Nereocaris*, *Canadaspis*, and other bivalved arthropods

Several Burgess Shale panarthropods have a bivalve carapace and were in the past associated with Crustacea, although they lack any crustacean synapomorphies. Legg and colleagues (2012) described a new genus, *Nereocaris* (Fig. 7.11d), and noted the presence of jointed biramus trunk appendages and segmental trunk sclerites. In their cladistic analysis, *Nereocaris* comes out as the basalmost member of a paraphyletic collection of bivalved stem euarthropods including such familiar forms as *Canadapsis*, that lies immediately crownwards of *Opabinia*.

7.4.4 Fuxianhuids: *Chengjiangocaris*, *Fuxianhuia*, and *Shankouia*

This series of Lower Cambrian Chengjiang fossils (Fig. 7.11c) represents a grade just basal to the euarthropodans having, according to Waloszek and colleagues (2005; Waloszek et al. 2007), evolved a number of new euarthropod characters. The head now consists of two segments, the first being the bearer of the compound eyes, and the second carrying a pair of grasping, uniramous antennae taken to be homologous with the single great appendage of the anomalocaridids. The second segment also carries a sclerotinized dorsal tergite that covers the next few segments dorsally and laterally. The trunk segments are well sclerotized and have softer, flexible membranes between the elements. Each bears a biramous limb comparable to those of *Nereocaris*, consisting of a segmented rod-like ventral part, the endopodite, plus a lateral, flap-like gill, the exopodite.

7.4.5 Megacheirans

Referred to as the 'short great appendage' arthropods, a series of Burgess Shale, Sirius Passet, and Chengjiang forms constitute the most derived euarthropod stem taxon, Megacheira (Fig. 7.12). They have a much shorter, more robustly built great appendage on the second segment, complete with a double peduncle at the base. In fact, not all authors agree that it is homologous with the anomalocaridid appendage, but that it is more likely with the labrum of the first segment found in most euarthropods (Budd and Telford 2009). Others believe that the appendage is actually the homologue of the chelicerae of the Chelicerata (Waloszek et al. 2005).

Three more segments, bearing biramous appendages, have been incorporated into a simple head shield, in which regard megacheirans are virtually completely euarthropodan. The trunk limbs also have an arthropodan structure, with fully distinct basipod and endopod, and a flat, paddle-shaped exopod (Stein 2010). Altogether, the megacheirans are structurally close to a hypothetical ancestral euarthropod (Waloszek et al. 2005; Waloszek et al. 2007). Another related genus, *Kiisortoqia* (Fig. 7.12c) from Sirius Passet (Stein 2010), possesses the four-segmented head, but its long, spiny, antennule-like great appendage more closely resembles those of anomalocaridids.

7.4.6 Conclusion: the evolution of arthropod characters

There continues to be debate and disagreement about the exact interrelationships of these and various other Cambrian taxa, both the more basal lobopodiforms and the various kinds of great appendage-bearing forms. Some authors have associated particular

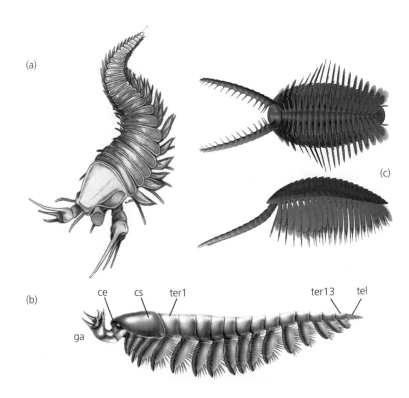

Figure 7.12 Megacheirans.
(a) *Kootenichela* (from Legg 2013);
(b) *Haikoucaris* (from Chen, J.-Y., et al. 2004a, reproduced with permission from John Wiley and Sons); (c) *Kiisortoqia* (from Stein 2010).

fossil taxa with specific euarthropod lineages. For example, both *Fuxianhuia* (Kühl et al. 2009) and the anomalocaridids (Chen et al. 2004a) have been proposed as stem chelicerates. Despite these differences on exact relationships amongst the various recent analyses (Budd 1996; Waloszek et al. 2005; Waloszek et al. 2007; Budd and Telford 2009; Edgecombe 2010; Legg et al. 2013; Edgecombe and Legg 2014), a general consensus has emerged in recent years about what the Cambrian fossil record reveals concerning the assembly of the euarthropod characters. Several broad grades in the transition from an ancestral 'lobopodan' or pan-arthropodan stage to a hypothetical ancestral euarthropod may be recognized. The first is that represented by *Opabinia*, the second by the 'great appendage' anomalocaridids, the third by the fuxianhuids. The fourth grade, represented by megacheirans, is very close to the ancestry of Euarthropoda.

Ancestral lobopodiform grade

Segmented and annulated body with a pair of simple, unjointed, non-sclerotized appendages on each segment.

Head more or less undifferentiated, consisting of only one or perhaps two segments.

Mouth simple and terminal.

Opabinian grade

Series of paired, segmental dorso-lateral folds or lobes above and a separate series of non-sclerotized ventral walking lobopods below.

Loss of the external annulations.

Head of two segments, the first bearing stalked, compound eyes, and the second bearing a single medial, flexible appendage with a pair of hook-like structures at the tip.

Mouth ventrally oriented.

Tail structure formed from the last few trunk segments.

Anomalocaridid grade

Fusion of the dorso-lateral folds with the lobopods to form pairs of segmental biramous appendages with gnathobase.

Sclerotization of the lobopods.

Euarthopodan-like gut diverticulum probably present by this stage.

Frontal appendages a pair of large feeding antennules divided to the base and composed of sclerotized, articulated podomeres.

Fuxianhuid grade

Trunk segments and lobopods well-sclerotized, with flexible membranes where the appendages arise.

Head still of two segments, but second segment bearing a sclerotized dorsal tergite extending backwards over the first few trunk segments.

Mouth posteriorly directed.

Megacheiran grade

Anterior three appendage-bearing trunk segments incorporated into the head. Antennule reduced in size.

7.5 Deuterostomia

The superphylum Deuterostomia was originally based on a series of embryological and larval characteristics, such as the pattern of early cleavage of the zygote and the secondary formation of the mouth during gastrulation. The taxon included the echinoderms, hemichordates, urochordates, cephalochordates, and vertebrates (craniates), and its monophyletic status has survived the molecular taxonomic revolution unscathed as the sister-group of Ecdysozoa plus Lophotrochozoa (Fig. 7.2). The long-disputed interrelationships amongst the individual deuterostome phyla have now been confidentially established, and there is strong agreement that the echinoderms and hemichordates are sister-groups, constituting Ambulacria (Swalla and Smith 2008). The remaining three continue to constitute the Chordata, but the longstanding belief that cephalochordates such as Amphioxus are the closest relatives of the vertebrates has been overturned. On the basis of molecular evidence it is accepted now that it is the tunicates, sea squirts, that occupy this position, and the two together form the clade Olfactores (Blair and Hedges 2005; Delsuc et al. 2006).

There are several putative stem members of Deuterostomia and its subtaxa in the Cambrian Lagerstätten, although in most cases disputes about their relationships continue because of differing interpretations of various ambiguous or poorly preserved soft structures, a problem exacerbated very considerably by the general paucity of derived anatomical characters shared amongst the different crown phyla themselves (Swalla and Smith 2008).

7.5.1 Vetulicolia: possible stem-group deuterostomes

Vetulicolia (Fig. 7.13a) and related forms are enigmatic Lower and Middle Cambrian fossils, common in the Chengjiang, Sirius Passet, and Emu Bay Shale faunas. They consist of a bilaterally symmetrical bipartite body. The large anterior carapace shows some indication of segmentation, and it is followed by a narrow, clearly segmented flexible tail terminating in a fin. Originally vetulicolians were interpreted as arthropods on account of what was interpreted as a bivalved carapace and intersegmental membranes between the tail segments, despite the absence of any sign of articulated appendages, or eyes. However, the one character on which authors agreed is that there is a line of filamentous gills associated with gill pouches and internal and external openings down each side of the carapace, a feature unknown in any arthropod, but characteristic of most extant deuterostome groups. There is also a structure that has been interpreted as a possible endostyle, the gutter in the floor of the pharynx that is only to be found amongst some deuterostome taxa. On these grounds, Shu and colleagues (2001, 2010) proposed that vetulicolians are basal deuterostomes. Aldridge and colleagues (2007) reviewed the group, and whilst noting the considerable difficulties surrounding interpretation of anatomical features, they tentatively agreed that a basal deuterostome is one of the more plausible possibilities. Others have suggested that vetulicolians are related specifically to tunicates within the Deuterostoma (Lacalli 2002; Garcia-Bellido et al. 2014) while Chen (2004) continued to maintain that they possess an exoskeleton and that their affinities lie closer to arthropods after all. Caron (2006) described the related form *Banffia*, but which lacks the critical character of gill slits, and he suggested it may be related to the protostomes rather than to the deuterostomes.

If ventulicolians are accepted as stem deuterostomes, then they imply that the ancestral grade before divergence of the crown groups was bilaterally symmetrical and segmented, and possessed a pharyngeal region of the alimentary canal with gill slits, and possibly an endostyle. The body surface was stiffened but lacked an exoskeleton.

7.5.2 Cambroernids: possible stem-group ambulacrians

Herpetogaster (Fig. 7.13b) is a Middle Cambrian Burgess Shale fossil restudied by Caron and colleagues (2010). It is a bilaterally symmetrical, stalked animal bearing a pair of branched tentacles on the head. There is a pair of pouched structures on either side of the head which they suggest represent gill pouches associated with pharyngeal openings. The body is clearly segmented and coiled somewhat, and the prominent stolon attaching to the substrate is pre-anal in position. A second, stratigraphically earlier form from Chengjiang is *Phlogites* (Xian-Guang et al. 2006). It is similar to *Herpetogaster*, although with a more robust stolon and tentacles, which suggests a more permanently sessile existence. These two genera, along with a number of other Cambrian forms such as *Eldonia* are included in a taxon Cambroernida. Referring to Swalla and Smith's (2008) conjecture that the common ancestor of the two ambulacrian groups, echinoderms and hemichordates, would have been a stalked, bilaterally symmetrical pharyngeal filter feeder with a paired water vascular system, Caron and colleagues (2010) suggest that the cambroernids are just such forms and therefore are stem ambulacrarians.

The implication of this hypothesis is that the ancestral ambulacrarian was not only stalked, but was also a segmental animal with tentacles, resembling a pterobranch hemichordate rather than an enteropneust hemichordate; the latter are inferred to be secondarily non-sessile. Cambroernids also support the idea of secondary loss of segmentation and bilateral symmetry in crown-group echinoderms. However, the warning needs to be reiterated in this case as much as in so many others that the interpretation of the homology of the preserved characters of the fossils is open to dispute. Most critically perhaps, the identification of the pair of small bulges on the head as true pharyngeal gill pouches is tentative at best. Indeed, Caron and colleagues'

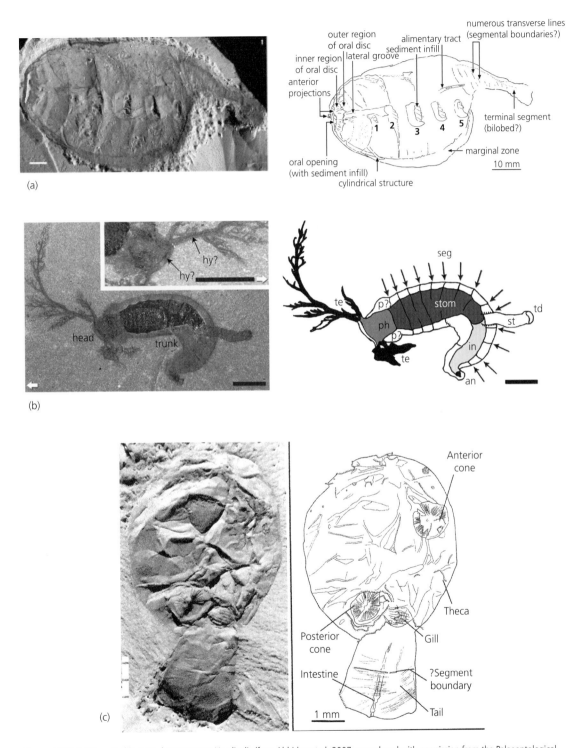

Figure 7.13 (a) The possible stem deuterostome *Vetulicolia* (from Aldridge et al. 2007, reproduced with permission from the Palaeontological Association and John Wiley & Sons); (b) the possible stem ambulacrian *Herpetogaster* (from Caron et al. 2010); (c) the possible stem echinoderm *Vetulocystis* (from Shu et al. 2010 reproduced with permission of the Royal Society).

(2010) conclusion is based as much on the negative observation that there is no convincing evidence that cambroernids are related to either the lopho-trochozoans or to ecdysozoans.

7.5.3 Vetulocystids: possible stem-group echinoderms

Vetulocystis (Fig. 7.13c) is another bipartite, ses-sile form from the Lower Cambrian Chengjiang of China, and as such has a certain similarity to the vetulicolians. It consists of a large theca, presum-ably representing the head, and a short but stout posterior tail which may be segmented. A possible intestine can be recognized running through the tail region. Three openings on the theca are interpreted by Shu and colleagues (2004) as, respectively, an an-teriorly situated mouth cone, a posteriorly situated anal cone near the theca–tail junction, and some kind of respiratory organ close to the anus. On the basis of comparison with the extinct asymmetrical homalozoan echinoderms (see next section), Shu and colleagues (2004) proposed that vetulocystids are basal echinoderms, the two groups sharing a re-duction of the gills and the geometrical relationship between mouth, respiratory organ, and anus.

However, as Smith (2004; Swalla and Smith 2008) has stressed, the morphological interpretations are very tentative indeed, and so far as yet known, vetulocystids do not possess a single unambiguous echinoderm character, neither the highly diagnos-tic histology of the stereom, the calcareous skeleton, nor a water vascular system. Shu and colleagues (2010, electronic supplement) are inclined to agree.

7.5.4 Carpoids: the asymmetric echinoderms

The crown-group echinoderms are characterized by a calcareous skeleton consisting of a unique histo-logical tissue called stereom. Each skeletal element is a single crystal of calcium carbonate forming a fine, spongy network. As such, fossil echinoderms are readily recognized and the fossils themselves are second only to vertebrates in the amount of biological information that potentially can be in-ferred from the skeleton alone. There is a series of entirely extinct echinoderms lacking the radial sym-metry characteristic of crown echinoderms that are

referred to as the carpoids. Carpoids are almost cer-tainly a paraphyletic group generally regarded as stem echinoderms on account of the stereom, but which have also been interpreted as stem mem-bers of other deuterostome groups, most notably of chordates. Jefferies (1986; Jefferies et al. 1996) went so far as to identify particular carpoid subtaxa as respective stem members of Cephalochordata, Uro-chordata, and Vertebrata, as well as of Chordata as a whole. However, accepting the proposed chordate relationships of carpoids depends on accepting a set of contentious homologies between various carpoid and chordate structures based only on the interpret-ation of impressions, foramina, and canals. What is referred to as the calcichordate theory also requires that the echinoderm carbonate skeleton consisting of stereom be lost three or four times independently, and in the case of the vertebrates replaced by the calcium-phosphate bony skeleton, a highly unpar-simonious proposition. A view of carpoid relation-ships at the other extreme has also been proposed, namely that they are highly derived crown-group echinoderms, probably crinoids, which have sec-ondarily lost radial symmetry (David et al. 2000).

The best known and also the most controver-sial carpoids are included in a taxon Stylophora (Smith 2005). They possess a large theca composed of skeletal plates, many small ones in the case of one of the two subtaxa, Cornuta (Fig. 7.14a), and a much smaller number of larger ones in the other, Mitrata (Fig. 7.14b). There is a single appendage which has been variously interpreted. In support of his calcichordate hypothesis, Jefferies (1986; Jef-feries et al. 1996) regarded it as the homologue of the chordate postanal tail, complete with evidence of a notochord, a dorsal nerve cord, and segmen-tally arranged myotomes. On the other hand, Da-vid and colleagues (2000) supported their crown echinoderm hypothesis by interpreting the append-age as an echinoderm ambulacrum or arm, claim-ing evidence for a water vascular canal and tube feet. Smith (2005), in his review, accepted the less egregious stem echinoderm hypothesis and accord-ingly accepted the structure as a muscle-bearing but non-segmental stem or stele, homologous to that of pterobranch hemichordates and the stalked crown echinoderms such as crinoids. To complete the range of views, Kolata and colleagues (1991)

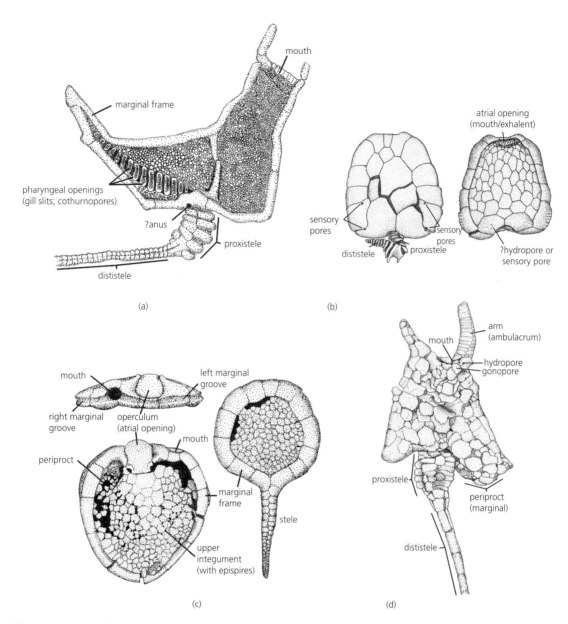

Figure 7.14 Carpoid echinoderms. (a) The cornute stylophoran *Cothumocystis*; (b) the mitrate stylophoran *Mitrocystites*; (c) the cinctan *Trochocystites*; (d) the solute *Dendrocystoides*. (From Smith 2005.)

thought that it was the homologue of none of these three, but a special organ, which they termed an aulacophore, suggesting that it was used for substrate or infaunal locomotion.

There are equally diverse views about the nature of the thecal anatomy. The boot-shaped cornutes have a series of external openings, but whether they are chordate-like gill slits or echinoderm-like respiratory structures is difficult to decide. To complicate matters, the mitrates actually lack these external openings, although they may have had internal gills opening into an atrium as in Amphioxus and tunicates, and therefore not indicated by the skeleton. Even the position of the mouth and anus are debated. The calcichordate hypothesis requires the anterior opening to be the mouth and the anus to be the opening near the attachment of the putative tail; the stem echinoderm hypothesis accepts the same opening as the mouth but possibly also including an opening for the anus or an atrium; the crown echinoderm hypothesis supposes it to be the periproct, effectively the anus, with the mouth located close to the appendage-as-ambulacrum. There are several more equally divergent opinions regarding such structures as a hydropore, and whether the presumed neural impressions at the bases of the theca indicate a chordate-like brain and cranial nerves, or an echinoderm-like aboral nerve centre and radial nerves.

Moving phylogenetically backwards, the cinctans (Fig. 7.14c) such as *Trochocystites*, are believed to be the most basal carpoids. They consist of a flat, superficially symmetrical theca with a single appendage which Smith (2005) interpreted as a posterior stele but which was reduced and did not attach to the substrate in life. There are three openings to the theca. One is at the front but offset to the right: it is interpreted as the mouth. On either side, marginal grooves lead from the mouth, the right one of which is less developed than the left. This may possibly indicate that a paired hydrovascular system was present, while the asymmetry may imply that the right hydrocoel was relatively reduced, indicating a more derived condition. The second aperture is situated on the upper surface of the theca, near the front but offset to the left side: it is taken to be the anus. The third aperture lies at the front edge of the theca, and Smith (2005) proposed that it is an atriopore leading from an atrium into which gills presumably opened from an enlarged, filter-feeding pharynx. Jefferies (1986) believed it to be a direct opening of a single gill to the outside.

The most derived of the carpoid subtaxa, Soluta (Fig. 7.14d), is of particular interest, since it is the most easily equated with echinoderms. They possess what is most reasonably interpreted as an ambulacrum similar to those of crown echinoderms in addition to an unmistakably echinoderm-like stele at the other end of the theca. The theca itself is highly asymmetrical. There is an apparent hydropore at the base of the ambulacrum, an oral opening presumably for the mouth and an aboral opening near the stele for the anus. There is no sgin

of gill openings either externally or internally. Thus a solute such as *Dendrocystoides* can more readily be interpreted as a stem echinoderm than can the stylophorans.

Smith (2005) reviewed the anatomy and phylogeny of the carpoids in the light of the molecular-based phylogenetic tree of the living deuterostomes. From the phylogeny he inferred that the echinoderm ancestor would have been bilaterally symmetrical, and possessed gill slits but no notochord. From the fossils he concluded that since the solutes are so echinoderm-like, and all the carpoids are related to one another, as a whole they should be interpreted as stem-group echinoderms. If this is accepted, then carpoids throw some light on the pattern and sequence of acquisition of echinoderm characters from an ambulacrian common ancestor, namely;

(i) Cinctan-grade: acquisition of the stereom skeleton.
(ii) Stylophoran-grade: loss of much of the bilateral symmetry, to be replaced by asymmetry.
(iii) Solutan-grade: loss of gill slits, acquisition of a water vascular system, and of a single arm or ambulacrum.
(iv) Crown echinoderm-grade: radial symmetry and multiplication of the ambulacra.

7.5.5 Crown Echinodermata

The crown-group Echinodermata is highly diverse, but in the great majority of cases there are no clear-cut stem members throwing light on how they evolved. One exception concerns the holothurians. Molecular evidence unambiguously supports a sister-group relationship between the echinoid and holothurian echinoderm classes (Pisani et al. 2012), and Smith and Reich (2013) described a basal holothurian *Palaeocucumaria* (Fig. 7.15) which is from the Early Devonian Hunsrück Slate Lagerstätte. It possesses radial water vascular canals with rows of tube feet, but these only extend a very short distance beyond the oral region, intermediate between the long rows found in echinoids and the greatly reduced rows of presumably homologous tentacles in crown holothurians. There are also reduced elements of the Aristotle's lantern of echinoids, and a stone canal with an external madreporite is still

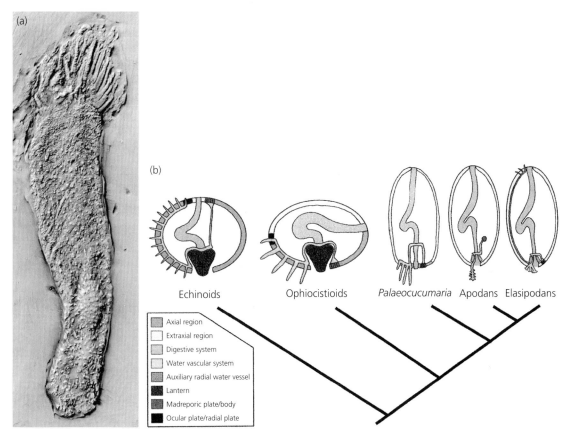

Figure 7.15 (a) *Palaeocucumaria*. (b) Sequence of stem holothurian grades. (After Smith and Reich 2013, reproduced with permission from John Wiley & Sons.)

present. The ophiocistioids such as *Gillocystis* are Devonian echinoid-shaped echinoderms that have reduced the water vascular and tube feet system to a degree intermediate between echinoids and *Palaeocucumaria*. Taking the series Crown echinoids, *Gillocystis*, *Palaeocucumaria*, and crown holothurians, Smith and Reich (2013) illustrated hypothetical stages in the transition to the very distinctive crown-holothurian body plan that involved modifications to body shape, proportions of the axial and extra-axial regions, reduction of the radial canals and tube feet, and reduction of the lantern.

7.5.6 Yunnanozoons: possible stem-group chordates

The Lower Cambrian Chinese genus *Yunnanozoon* (Fig. 7.16a) and the probably synonymous *Haikouella* is an elongated, fish-shaped animal with a gill-bearing head region and a muscular trunk terminating in a tail. All commentators have agreed that it is a deuterostome, but its relationships within the superphylum continues to be the subject of one of the hottest disputes in Cambrian invertebrate phylogeny. Initially *Yunnanozoon* was interpreted as a cephalochordate on the grounds of a large pharynx with branchial arches and endostyle, segmental musculature, notochord, and segmental gonads (Chen et al. 1995), although shortly afterwards it was reinterpreted as a hemichordate on the grounds of a supposed tripartite body and a proboscis (Shu et al. 1996). However, the issue seemed settled in favour of stem-vertebrate membership when much better preserved material was described as being encephalized, with a large brain and small eyes, pharyngeal gill bars and gill filaments, and importantly in possessing myomeres rather than simple segments. However, Shu and colleagues (Shu et al. 2003, 2010) seriously disputed these anatomical interpretations. They found no evidence for myomeres, only for non-overlapping segmental blocks, and nor could they identify eyes, nostrils, branchial bars, or a notochord. They did agree that pharyngeal gills are present, and therefore that yunnanozoons are deuterostomes, but that they are at the base of that clade and possibly related to vetulicolians. Indeed, they speculate that the yunnanozoon body form was derived from that of a vetulicolian-like form by the posterior segmental region evolving an anterior extension over the gill-bearing anterior region of the body. Other reviewers continued to support the chordate or near-chordate relationships of

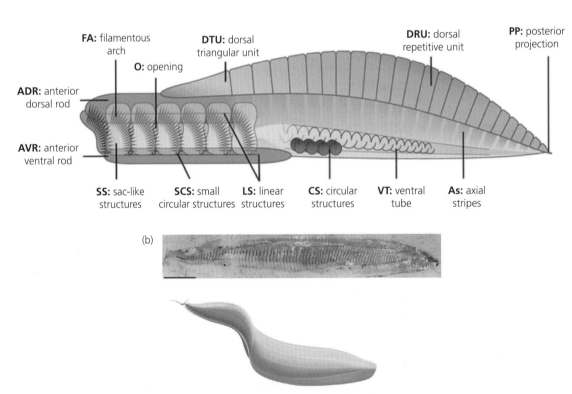

Figure 7.16 (a) *Yunnanozoon* (from Cong et al. 2014, reproduced with permission from the Palaeontological Association and John Wiley & Sons); (b) *Pikaia* (from Conway Morris and Caron 2012, reproduced with permission from John Wiley & Sons); (c) the conodont *Clydagnathus* (from Briggs et al. 1983, reproduced with permission from John Wiley & Sons); (d) *Haikouichthys* (from Shu et al. 1996, reproduced with permission from Nature Publishing Group).

(c)

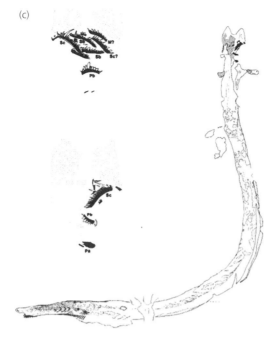

(d)

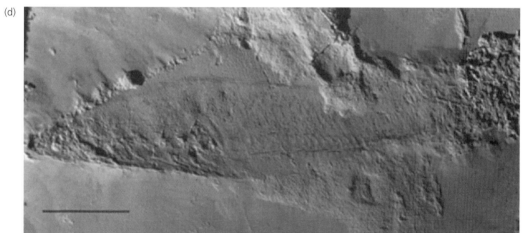

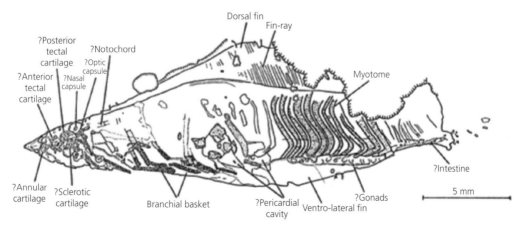

Figure 7.16 (*continued*)

yunnanozoons (Swalla and Smith 2008; Mallat and Holland 2013), while at the other extreme, Cong and colleagues (2014) studied several hundred specimens, and still concluded that no more than a bilaterian assignment can be confidently claimed. Unless and until their homologies are satisfactorily resolved, yunnanozoons must be judged most unhelpful for understanding chordate evolution and the pattern of acquisition of their characters (Donoghue and Keating 2014).

7.5.7 *Pikaea*: possible stem-group chordate

Originally interpreted as an annelid, the Burgess Shale animal *Pikaia* (Fig. 7.16b) was reinterpreted by Conway Morris (1977) as a basal chordate and fully described in 2012 (Conway Morris and Caron 2012) with the same conclusion. It is a small, fish-shaped organism varying in length from 1.5–6 centimetres and bearing a dorsal and a ventral median fin and a pair of tentacles on the head. On either side of the head region there is a series of small appendages whose identity and function are difficult to interpret; there is some slight indication they may be associated with pharyngeal pores and therefore are gills. The only chordate characters identified with confidence are metameric organization of the body, a nerve cord and a notochord, and a possible ventral blood vessel. In an extensive subsequent commentary, Mallat and Holland (2013) doubted the identification of the notochord, and noted the evident absence of a brain, endostyle, and postanal tail, implying that *Pikaia* is the most basal chordate and furthermore that it is actually highly derived compared to amphioxus, which makes the latter a better model for a pre-vertebrate chordate stage. Indeed, also in contrast to Conway Morris and Caron (2012), they considered that yunnanozoons are closer than *Pikaea* to the vertebrates.

7.5.8 Conodonts: possible stem-group vertebrates

The mystery of the phosphatic, tooth-like elements that constitute the conodonts appeared to have been largely solved after a great many years by the discovery of soft-bodied, worm-like animals in which conodont tooth plates are found: they were segmented

with V-shaped myomeres and also possessed fins with fin rays (Fig. 7.16c). There is also some evidence for the presence of paired eyes and possibly a notochord. The conodont teeth lie in the head region and are therefore assumed to be vertebrate teeth, although no details of branchial arches or possible jaws were evident in the specimens (Briggs et al. 1983; Aldridge et al. 1986; Aldridge et al. 1993). While this interpretation is widespread, others have seriously questioned it, as reviewed at length by Turner and colleagues (2010), who argued that the absence of significant cephalization, the V-shaped rather than W-shaped myomeres, and the absence of any trace of dermal or endo skeleton point to a decidedly pre-vertebrate grade of organism. The histology of the teeth is also unlike vertebrate hard tissues in several respects. They concluded that at most conodonts are stem-group chordates, and a relationship of at least some of them with an entirely different phylum, Chaetognatha, cannot really be excluded. In any event, as with the yunnanozoons, conodonts add little to an account of the origin of vertebrate characters (Donoghue and Keating 2014) as long as interpretations range from possessing almost none to possessing almost all the characters of crown vertebrates.

7.5.9 *Myllokunmingia*, *Haikouichthys*, and *Metaspriggina*: stem-group vertebrates

There are a number of undoubted stem-group vertebrates from the Early Cambrian of Chengjiang and the Middle Cambrian Burgess Shale. Although Lagersättten forms, they are amongst the least contested of the Cambrian soft-bodies taxa, and exhibit such characteristic vertebrate features as paired nasal sacs and camera-type eyes, pouched gills, and evidently a heart. The Chinese *Myllokunmingia* (Xian-Guang et al. 2002), and the possibly synonymous *Haikouichthys*, are small, fish-shaped animals a few centimetres in length, possessing dorsal and ventral fins but lacking fin rays (Fig. 7.16d). The body is built of myomeres, although they are V-shaped as in amphioxus rather than W-shaped as in vertebrates. Another amphioxus feature is the presence of a series of segmentally arranged structures presumed to be gonads. About six gill pouches surrounded by a branchial basket and filamentous gills occupy the anterior region, and these are followed by at least 25 more arches behind, as in larval amphioxus. There is a terminal mouth and behind it are two pairs of small structures interpreted as optic and otic capsules respectively, which implies a degree of vertebrate-like cephalization.

Metaspriggina (Conway Morris and Caron 2014) is a similar Canadian form but it differs in having myomeres tending to be W-shaped, suggesting that, as befits their younger age, they are closer to the crown vertebrates.

CHAPTER 8

The vertebrate fossil record

The vertebrate fossil record is far more informative than the invertebrate record about the pattern of acquisition of characters along lineages culminating in new higher taxa for several reasons. The internal calcium-phosphate skeleton is more complex and therefore homologies and inferred soft structures are more easily determined, as are detailed aspects of the functional anatomy and adaptive biology of the organism. Furthermore, the major taxa of vertebrates originated a good deal more recently than the equivalent invertebrate phyla and classes, and this is especially true of the terrestrial taxa, which helps to explain why a greater number of intermediate-grade fossils have been described. As a consequence, most of what may be inferred from the fossil record about how the derived traits of a high-level taxon are assembled in evolutionary time is based on the vertebrates. Amongst the vertebrate taxa in this regard, particularly informative cases are the tetrapods, birds and mammals. Significant stem-grade fossils of a few other taxa also throw some light on their evolutionary trajectories, notably turtles and, within Mammalia, the whales. These are the only taxa that are discussed; other higher vertebrate taxa for which there is much less in the way of significant sequences of intermediate grade fossils are not dealt with here.

The phylogenetic interrelationships of the main vertebrate taxa (Fig. 8.1) are on the whole uncontroversial, apart from a few points of issue (Meyer and Zardoya 2003; Hedges and Kumar 2009; Fong et al. 2012). One problem is the base of the living vertebrate groups, and the long-standing morphological-based dispute about whether the two cyclostome groups, lampreys and hagfishes, are monophyletic or whether the lampreys are related to the gnathostomes. This has been largely settled by molecular evidence in favour of the former view, though not without some dissent (Near 2009). A second still somewhat controversial case is the relationship amongst the major lobed-fin fish taxa and the tetrapods. The molecular evidence now points to the Dipnoi being the sister-group of tetrapods, rather than the coelacanth (Liang et al. 2013). Amongst the tetrapod taxa, there are two outstanding issues (Fong et al. 2012). One concerns the interrelationships of the three modern amphibian groups Anura, Apoda, and Urodela. With a number of variations, the morphological evidence has tended to support the view that the modern amphibians are paraphyletic, with the apodans related to the tetrapods (Carroll 2009). Molecular evidence, however, has tended to support a monophyletic lissamphibia containing all three amphibian groups, although with no agreement on how the three are themselves interrelated. The second current issue in tetrapod phylogenetics concerns the position of the turtles, which are considered by different authors to be the sister-group of crocodiles, of archosaurs, of squamates, of sauropsids, and of all other amniotes (Fong et al. 2012).

The difficulty of resolving these questions creates the question of why this should be so. It suggests that the situation in each case is actually a 'soft' polytomy, which is where the divergence of the respective taxa from one another was so rapid, or to be more precise so closely spaced within both morphospace and molecular space that what may in reality have been successive dichotomous splits are indistinguishable evidentially from a simultaneous polytomy.

8.1 Mammals

The higher taxon whose origin is most comprehensively illustrated by the fossil record are mammals,

The Origin of Higher Taxa, First Edition. T. S. Kemp.
© T. S. Kemp 2016. Published 2016 by Oxford University Press and The University of Chicago Press.

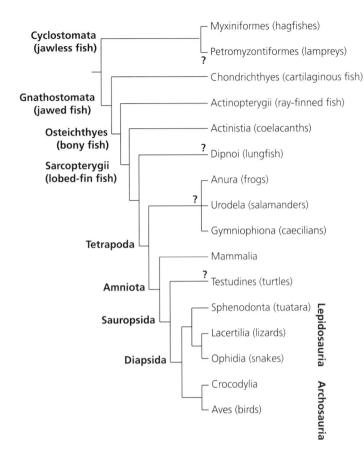

Figure 8.1 Phylogeny of the vertebrates.

which have received the most focused if not the most voluminous attention as a paradigm for evolution at this level (Kemp 1982; Kielan-Jaworowska et al. 2004; Kemp 2005). There are around ten stem-group grades of fossil taxa, between the hypothetical common ancestor of mammals and reptiles at the base and the ancestral crown group Mammalia at the termination of the lineage, each with a different combination of ancestral and derived characters (Fig. 8.2). Collectively constituting the non-mammalian Synapsida ('mammal-like reptiles' in informal language), their earliest appearance is in the Upper Carboniferous, while the first mammaliaforms occur in the Upper Triassic. This is a span of about 120 million years, which gives a rough indication of how long it took to evolve all the evolutionary changes that took place in the transition from a scaly, ectothermic, sprawling-limbed, small-brained, and simple-toothed amniote to a fur-bearing, endothermic, parasagittally gaited,

large-brained, complex-toothed, and behaviourally complex mammal. The interrelationships amongst the main lineages of synapsids are generally agreed upon (Fig. 8.2), although there are arguments about the cladistic position of some of the more derived taxa (Kemp 2009a, 2012). The most basal forms have traditionally been referred to as 'pelycosaurs' (Fig. 8.3) which, although a paraphyletic assemblage, does draw attention to one of the largest morphological gaps in the series, namely that occuring between this grade and all other synapsids. The latter are members of the more derived clade Therapsida.

8.1.1 The grades of fossil stem mammals

Basal pelycosaur-grade: Eothyrididae

Only isolated skulls, one of *Eothyris* (Fig. 8.3) and one plus some other fragments of *Oedaleops*, are

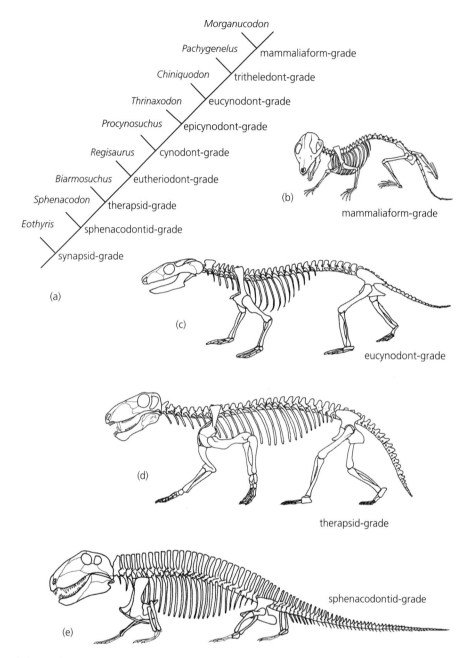

Figure 8.2 Cladogram of representative genera of the principal stem-mammal taxa known, with reconstructions of the skeletons of some. (Modified after Kemp 2007a.)

known so far of this important taxon, and they are from the Lower Permian of North America, close to the Permo-Carboniferous boundary, which is actually somewhat later than the first appearance of synapsids in the Upper Carboniferous (Reisz et al. 2009). Nevertheless, in all cladistic analyses eothyridids emerges as the most basal synapsid clade. Plesiomorphic features include a broad, low skull shape and a large supratemporal bone at the rear of the skull roof that are the ancestral conditions for

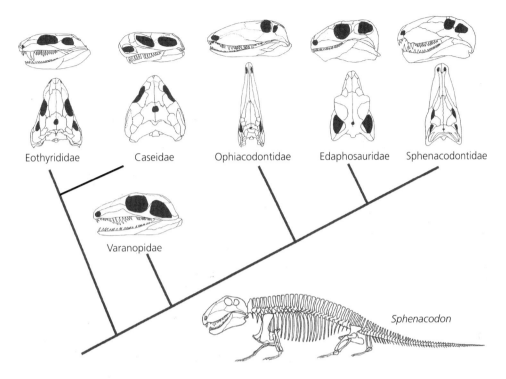

Figure 8.3 Phylogeny of pelycosaur-grade synapsids. (Modified after Kemp 1988.)

amniotes. The main derived mammalian character they possess is a synapsid temporal fenestra, a space in the side of the skull low down behind the orbit, which is the most prominent diagnostic character of the synapsids. There is also an enlarged tooth or teeth in the upper canine region of the maxilla.

Mention should also be made of the related caseid pelycosaurs which make their appearance significantly later in the Lower Permian. They are highly specialized for a herbivorous mode of life such as having a short, strongly built skull, leaf-like teeth, massive jaw-closing muscles, and relatively large, barrel-shaped bodies. As such, they add no further information about the acquisition of derived mammalian characters.

Basal eupelycosaur-grade: Varanopidae and Ophiacodontidae

The skull of a varanopid such as *Varanops* (Fig. 8.3) is a good model for the ancestral condition of the rest of the pelycosaur-grade synapsids, the 'Eupelycosauria'. It has a higher, narrower shape, elongated snout, and larger temporal fenestra, while the supratemporal bone is reduced to a splint. The dentition consists of very sharp, somewhat recurved teeth that vary gradually in size along the tooth-row with those in the middle of the row the largest. The ophiacodontids are generally quite similar although with elongated skulls, and they include the earliest known pelycosaur skeleton, *Archaeothyris* from Upper Carboniferous deposits of Nova Scotia. The postcranial skeleton of all pelycosaur-grade animals (Fig. 8.2) is still generally primitive, with heavily built limbs, a massive, fairly immobile shoulder girdle, restricted range of fore-limb movements, and a small ilium and large pubo-ischium of the pelvis. The limbs were held in the sprawling mode of typical reptiles, and the digits were of different lengths, the fourth being longest. However, they do possess one notable incipiently mammalian character of the postcranial skeleton, namely the obliquely inclined rather than horizontal zygapophyses of the dorsal vertebrae, which indicates a reduction of the lateral undulatory component of locomotion that is characteristic of more basal amniotes.

Eupelycosaur-grade: Sphenacodontidae

Dimetrodon (Fig. 8.3) is the most familiar pelycosaur. As a member of the carnivorous family Sphenacodontidae, it is characterized by distinct enlargement of caniniform teeth compared to the rest. The temporal fenestra is relatively large and the coronoid region of the lower jaw broadened, which indicates a significant increase in the jaw adductor musculature. There is also a ventro-lateral keel on the anglar bone, which extends backwards lateral to the main body of the jaw as a small reflected lamina. This, a feature exaggerated in the subsequent therapsids, is associated with elaboration of the more ventral jaw musculature.

Sphenacodontids lived from the Upper Carboniferous through the Lower Permian and over this time were the dominant carnivorous terrestrial tetrapods. As were all pelycosaur-grade synapsids, they were sprawling-limbed, simple-toothed animals. The histology of the bone points to relatively slow, seasonally interrupted growth and they undoubtedly possessed a low metabolic rate and ectothermic physiology. Their habitat so far as yet known was restricted to the hot, humid equatorial regions of the Upper Carboniferous and Lower Permian of Pangaea (Kemp 2006).

Biarmosuchia

The most basal therapsid so far found is believed to be the Middle Permian Chinese form *Rarianimus*, but so far it is represented only by the anterior half of a poorly preserved skull (Liu et al. 2009) which adds little to understanding the origin of therapsids for all its potential importance. Otherwise, the probably slightly younger biarmosuchians (Fig. 8.4) are the most basal therapsids. They are related to the sphenacodontid 'pelycosaurs' as indicated by the reflected lamina of the angular bone of the lower jaw, which is small in the latter but in therapsids has become a large, thin sheet of bone that lies lateral to the body of the bone, leaving a very narrow space between. The primary function of the reflected lamina is assumed to have been for mandibular muscle attachment, as indicated by strengthening ridges across the surface, and was probably associated with backward-running musculature for jaw opening. Biarmosuchians share other characters with sphenacodontids, notably the enlarged, though by this stage very much enlarged, upper and lower caniniform teeth and differentiated incisors, an elevated coronoid eminence of the lower jaw, and a characteristic strengthening of the occipital region of the back of the skull. There are also a few characters of the postcranial skeleton shared between the two taxa, especially a degree of narrowing of the scapular blade. However, therapsids

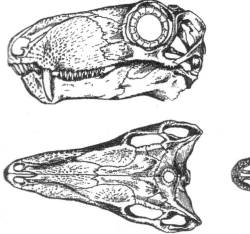

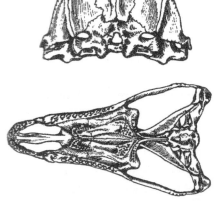

Figure 8.4 *Biarmosuchus.* (From Ivakhnenko 1999.)

also have a great many new, derived characters that taken together imply a significant shift in biology and physiology (Fig. 8.2). The temporal fenestra is much larger and the associated adductor musculature therefore presumed to have been more developed, whilst the jaw is relatively shorter, which had the effect of increasing the bite force at the teeth. There is a pronounced broad trough in the roof of the mouth, the choanal trough, which would have helped breathing to continue whilst feeding. In the postcranial skeleton, the limbs are longer and more gracile. The shoulder girdle is much more slender and capable of shifting and rotating on the ribcage, while the anatomy of the shoulder joint evolved to allow a shift from a highly predetermined and limited pattern of movement to a much freer mode permitting greater range and amplitude of forearm movements. The tail is reduced in size and the ilium extended, indicating a tendency to shift the hind-limb retractor musculature from the caudal vertebrae to the pelvis. Therapsid bone histology indicates a shift from the slow-growing, poorly ossified lamellar-zonal bone of typical ectotherms to the faster growing, well-vascularized fibro-lamellar bone typical of most mammals (Chinsamy-Turan 2012a). Whether bone histology is a reliable indicator of endothermic temperature regulation as well as growth rate is disputed. However, taking all these evolutionary changes into account, it is virtually certain that the most significant general biological shift from 'pelycosaur' grade to therapsids was towards partial endothermy and the implied increase in the range of ambient conditions in which the animals could remain active (Kemp 2006, 2006a).

Therocephalia

Virtually at the same time as the first appearance of therapsids in the Middle Permian fossil record, several more derived therapsid taxa occur, of which the therocephalians (Fig. 8.5) represent the next grade on the hypothetical lineage leading towards mammals. Here the temporal fenestra has enlarged by expanding inwards, resulting in a pair of broad, almost horizontal fenestrae separated by a narrow, midline girder, the sagittal crest. The postcranial skeleton has also acquired several incipiently mammalian characters. The dorsal vertebrae consist of

a thoracic region with long ribs forming a ribcage, and a posterior lumbar region in which the ribs are shorter, stouter, and more horizontally oriented. In the pelvis, the ilium has expended forwards somewhat, in association with an additional trochanter on the femur, the homologue of the mammalian trochanter minor. The ankle has evolved a moveable joint between the astragalus and calcaneum, which is the homologue of a joint between the same two bones in mammals (Kemp 1978), and the digits of the feet are approximately the same length as one another, again as in mammals.

Procynosuchia

The next grade of synapsids represented in the fossil record are the cynodonts, amongst which there are actually several increasingly mammal-like taxa known. The most basal forms are the procynosuchians (Fig. 8.6a), notably the Upper Permian *Procynosuchus* (Kemp 1979, 1980a). New mammalian characters of the skull include the dentition, which is differentiated into incisors, canines, slightly expanded premolariforms, and molariform teeth bearing a ring of cuspules around the base of the crown. Upper and lower molariforms operated against one another, although at this stage not in the precise occlusion found in mammals. The adductor jaw musculature is more markedly mammalian than that of the therocephalians, as indicated by the deeper sagittal crest, and the expansion of the muscle origin along the lateral bar of the temporal fenestra, known as the zygomatic arch. This is the start of the differentiation of the adductor muscle into distinct medial temporalis and lateral masseter muscles (Fig. 8.8). The lower jaw accommodated these changes by evolving a broad coronoid process on the upper part of the dentary, and a masseteric fossa on its lateral face. Indeed, the dentary as a whole is relatively enlarged and the remaining jaw bones, collectively referred to as the postdentaries reduce. The jaw hinge bones too, quadrate above and articular below, are reduced in size as is the reflected lamina of the angular. It is clear that this part of the mandible had become involved in the reception of airborne sound waves, a function increasingly manifested in later cynodonts and mammals as described later in this chapter (Fig. 8.9a). Another innovative feature of the procynosuchian skull is the

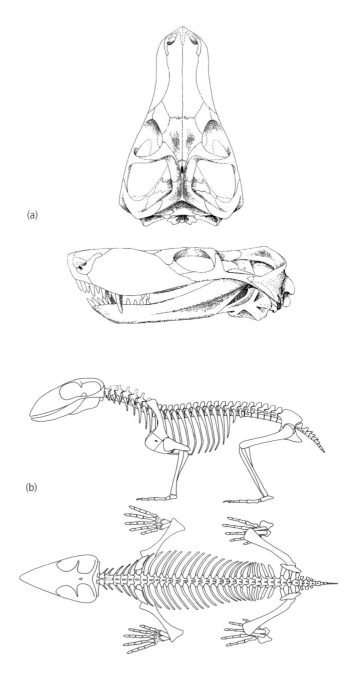

(a)

(b)

Figure 8.5 Therocephalians. (a) Skull of *Oliveria*. (b) Skeleton of *Regisaurus*. ((a) from Brink 1965; (b) from Kemp 1986.)

secondary palate in the roof of the mouth, which is incomplete medially at this stage although presumably completed in life by soft tissue and therefore able to perform the function of completely separating a respiratory air passage above from the masticating food cavity of the mouth below. The procynosuchian postcranial skeleton is not greatly modified from the therocephalian grade, although more mammalian than therocephalians in having the first two vertebrae, atlas and axis, modified to increase head movements, and a more mammalian pelvis and hind-limb structure and function.

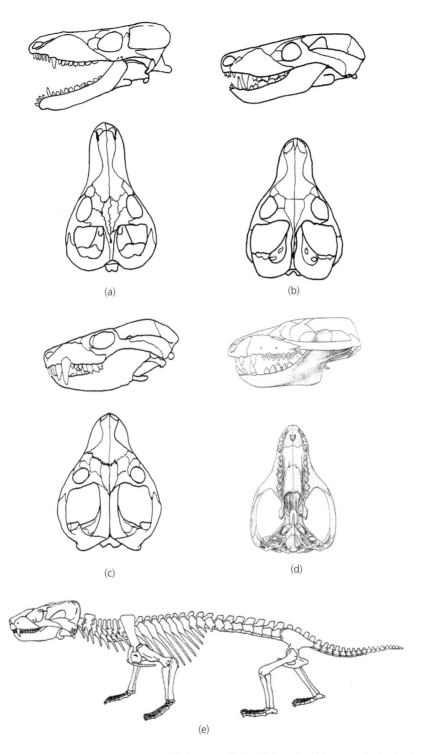

Figure 8.6 Cynodonts. (a) The procynosuchian *Procynosuchus*; (b) the epicynodontian *Thrinaxodon*; (c) the eucynodontian *Lumkuia*; (d) the tritheledontian *Pachygenelus*; (e) skeleton of *Thrinaxodon*. (From Kemp 2012, redrawn from various sources.)

Epicynodontia

The most familiar and abundant epicynodont-grade cynodont known is *Thrinaxodon* (Fig. 8.6b and e), which was a relatively small Lower Triassic form. It was more derived than procynosuchians in a number of features. The adductor musculature of the jaw was further enlarged as indicated by the deeper sagittal crest and larger coronoid process for the temporalis component, and the more massive zygomatic arch and broader adductor fossa of the dentary for the masseter (Fig. 8.8c). The dentary was relatively larger, and the postdentary bones, quadrate, and also the reflected lamina smaller still than in procynosuchians. In the postcranial skeleton, the limbs are more slender and elongated, and the hind limb by this grade had adopted an obligatory parasagittal gait.

Eucynodontia

The third clearly defined cynodont grade is Eucynodontia (Figs 8.2c and 8.6c), in which the dentary is by far the dominant bone of the lower jaw, and the postdentary bones reduced to a slender rod attached to a trough in the medial side of the hind part of the dentary. There is still a jaw hinge between the quadrate and the articular, although it is small and could have transmitted no more than small stresses, and by this stage was part of a sound reception system functioning basically like that of mammals, although probably without an air-filled tympanic cavity. The jaw musculature itself had achieved an effectively mammalian grade, producing a very high bite force concentrated at the teeth and very little reaction at the jaw hinge. At the same time the reorganization led to an antero-posterior and latero-medial balance of the forces applied to the jaw and so the possibility of very precise movement of the mandible in the horizontal plane. This latter property correlates with the evolution of both enlargement of and precise occlusion between the upper and lower postcanine teeth.

The size of the brain in cynodonts is impossible to judge accurately because, as in non-mammalian and non-avian amniotes generally, it did not fill the cranial cavity. However, the dimensions of the cavity do indicate that eucynodontians had evolved an enlarged cerebellum and probably reduced midbrain (Fig. 8.9f) compared to more basal forms (Kemp 2009).

The postcranial skeleton of eucynodonts also shows a significant transition towards the mammalian condition, particularly in the structure and inferred function of the hind limb. The ilium is elongated forwards and the pubis and ischium reduced and rotated backwards, and there are mammal-like major and minor trochanters on the femur, indicating hind-limb musculature of essentially the mammalian pattern.

Tritheledontia

A number of mainly very small, sharp-toothed forms from the Upper Triassic of South America and South Africa are rather loosely grouped as tritheledontians (Fig. 8.6d). Their interrelationships have not yet been firmly established, and it is probably a paraphyletic assemblage with some forms closely related to mammaliaforms and others more divergent (Bonaparte et al. 2003). They are of small body size, with a skull length of around 30–60 mm, and anatomically are at least superficially very mammal-like. There is a simple contact between the dentary bone of the jaw and the squamosal of the skull and they also share with the mammaliaforms loss of the postorbital bar, a similar tooth action, prismatic enamel in the teeth, and a postcranial skeleton that is almost indistinguishable in broad features from that of a mammaliaform such as *Morganucodon*.

Mammaliaformes

By the end of the Triassic, small, extremely mammal-like synapsids had appeared, such as the worldwide Morganucodontia (Fig. 8.7). Their most eminent characteristic is the presence of a fully developed new jaw articulation between a condyle on the dentary and a glenoid fossa in the squamosal. This morphology correlates with the mode of action of the dentition, in which the lower postcanine teeth move inwards as well as upwards to occlude accurately with the upper postcanines. There was also a further reduction of the original jaw articulation bones, quadrate and articular, although they were still attached to the cranium and jaw respectively,

Figure 8.7 Skeleton of the basal mammaliaform *Morganucodon* (from Jenkins and Parrington 1976, reproduced with permission from the Royal Society).

indicating an ability to detect higher-frequency sound. The brain is substantially larger relative to body size than in non-mammalian cynodonts, and there is a significantly enlarged forebrain suggesting that a primitive neocortex had evolved.

Only a few minor plesiomorphic characters distinguish the postcranial skeleton from that of a crown-group mammal, and the high level of agility of locomotion that is found in small living mammals must have evolved by this stage.

8.1.2 The origin of the mammalian body plan

As just briefly reviewed, the palaeontological evidence bearing on the origin of mammals embraces the most complete sequence of intermediate grades of the entire fossil record of the origin of higher taxa. There is also the advantage shared with other vertebrate taxa that several aspects of the functional biology and physiology of mammals is reflected to a remarkable extent in skeletal morphology. This particular case is therefore an important paradigm for investigating major, complex evolutionary transition in general (Kemp 1982, 2005, 2007a). Over the course of about 120 Ma, the lineage from basal amniote to mammal underwent considerable evolutionary changes in dentition and jaw mechanism, gait and locomotory performance, heart and circulatory anatomy, precise endocrine-controlled chemoregulation of body fluids, sensitivity of olfaction and hearing,

brain size and elaboration of its functional capability, parental provision of the young, and all the physiological processes and functions associated with the transition from ectothermy to endothermy.

All these structures and processes are integrated with one another as parts of the complex system that is the overall biology and life of the mammal (Fig. 8.10). However, for analytical purposes it is helpful to begin by considering one aspect of the system as central to the rest, and the best candidate for this role is endothermy, to which virtually every aspect of mammalian biology is related. Many of the diagnostic mammalian characters, such as elevated cell mitochondrial number, hair, and the double circulation of the heart and vascular system, are parts of the process causing the several-fold increase in metabolic rate and high body temperature. Other characters are associated with activities that endothermy facilitates, including nocturnality, high sustainable aerobic activity levels, and rapid juvenile growth.

Endothermy

Endothermic physiology simultaneously serves several functions in modern mammals, and attempts to decide which one is in some sense primary are fruitless (Kemp 2006a). This is because the functions are inevitably integrated with one another by shared physiological mechanisms and ecological consequences (Fig. 3.6), and also because the extent and relative importance of each function differs in detail from species to species (Lovegrove 2011). Actually there are two broad adaptive categories. First, the relatively precise thermoregulation that maintains the body temperature to within about 2 °C for most of the life of most species; second, the elevated maximum aerobic metabolic rate, typically between five and fifteen times that of similar sized ectotherms. These two physiological properties together opened up a radically new mode of life, based on high-energy expenditure. The ability to regulate the internal body temperature greatly increased the range of ambient temperatures, elevated as well as reduced, within which active life is possible. Initially this may have been an adaptation to the cool conditions of nocturnal life, but it also permitted extension of the habitat into more seasonal climates. A constant internal temperature also has physiological as well as ecological consequences by promoting more constant rates of the many interlinked temperature-sensitive processes in the body, such as enzyme activity, diffusion of transmitter molecules across synapses, and speed of muscle fibre contraction. This rate constancy is necessary for the correct and reliable functioning of the complex system, as manifest by the entire organism's biology, and is a *sine qua non* for the integrated activity of the hugely enlarged mammalian central nervous system, and its associated behaviour patterns. The second aspect of endothermy, the elevated aerobic metabolic rate, endows mammals with a much higher maximum level of sustainable muscle activity as manifest in such behaviours as hunting, predator avoidance, rate of food collection for the juvenile, dispersal, and mating behaviour.

Endothermy is a far from simple system (Fig. 3.6). At the cellular and molecular level, it requires a large increase in metabolic heat production, with a basal (resting) metabolic rate five to ten times that of a comparable sized ectotherm and a maximum aerobic rate increased by a similar ratio. This involved the evolution of increased numbers of larger mitochondria in the cells, and a shift in their biochemical activity to produce more heat directly, in place of ATP. But increased mitochondrial heat production alone is not enough. There must also be the means to balance heat production and heat loss, which requires as a minimum detection of body temperature, hormonal regulation of metabolic rate, variation in skin conductivity, finely controllable vascularization of blood flow to the body surface and oral cavity for evaporative cooling, and a whole variety of behavioural reactions, daily and seasonal, all precisely integrated by the interaction of the somatic and autonomic nervous systems. Beyond these, there are structures and mechanisms that evolved for enhanced ventilation rates, both resting and even more extremely during very high levels of aerobic muscular activity. To fuel the system, the rate of food assimilation needs to match the increase in metabolic rates, with implications for the sensory physiology of food detection, the mechanics of food collection and intake, the process of assimilation, and the need for large storage reserves in the body. All these activities and processes require an increased level of neuromuscular control. A final essential capacity for endothermy is parental provision for the juveniles, because it takes an increased viable body size and amount of growth time for the full system to develop, during which the juvenile is unable to regulate its own internal environment or indeed function as an independent organism at all: a regulated external environment must therefore be provided by the parent during development. This may be by means of external incubation of embryos in eggs in a nest or burrow, as in modern monotremes and presumably basal mammals generally, or by an extended period of intrauterine gestation as in modern therians, in both cases followed by lactation.

Direct, incontrovertible evidence for the appearance of endothermy in non-mammalian synapsids is largely lacking in the fossil record since a keratin-insulating layer of hair is unlikely to be preserved. The closest thing is the claim that maxillo-turbinates occur in the nasal cavity of therocephalians and cynodonts (Hillenius 1994; Ruben et al. 2012). In mammals, these structures, which lie in the path of the air flow to and from the lung, function to warm and humidify the inspired air, and to recapture water by condensation from the expired air. The presence of fine ridges on the internal surface of the snout have been interpreted as the site of attachment of maxillo-turbinates, and Laass and colleagues (2010) have found actual sheets of preserved cartilaginous tissue in the specialized herbivorous therapsid *Lystrosaurus*, which is certainly very suggestive. A more comprehensive argument for the origin of endothermy is presented shortly, after first discussing other biological attributes that evolved in mammals for which there is often very clear morphological evidence from fossil specimens.

Feeding

To an extent unique amongst animals, mammals evolved a jaw action that reconciled the normally incompatible properties of a very large muscular force and fine precision of mandible movement (Kemp 2005). The dentition evolved from the simple-toothed homodont dentition of basal amniotes to the mammalian dentition of functional differentiation of tooth types that separates food-collecting incisors and canines at the front from multi-cusped occluding molariform teeth for mastication at the rear. To generate the huge bite force that is required for tooth-to-tooth mastication, the adductor mandibuli musculature enlarged. Its attachment to the skull occupies the edges and connective tissue cover of the enormous temporal fenestra, and its area of insertion on the mandible is greatly increased by the coronoid process and the broad angular region of the dentary (Fig. 8.8). Furthermore, the geometry of the two major components of the adductor mandibuli, the temporalis internally and the masseter externally, are altered so as to concentrate virtually all the net force at the point of bite. On top of this, the horizontal vectors of the muscle forces in the horizontal plane are more or less balanced, which permits the necessary level of fine control in this plane for precise occlusion of lower and upper teeth. Clearly, the evolution of the mammalian system of feeding required correlated changes in dentition, jaw shape, muscle anatomy, oral cavity food manipulation by tongue and lips, and of course neuromuscular control of the entire system. Within the fossil record of synapsid grades, much of this transition can be followed directly, or readily inferred from the anatomy.

Ventilation

There are several skeletal correlates of a significantly enhanced ventilation capacity, all of which can be seen within the synapsids. The earliest of these to evolve is the more progressive gait (Fig. 8.2), in which the limbs, initially the hind limb, approach a parasagittal orientation so that the feet are brought under the body. This potentially increases the lung volume by raising the ribcage off the ground and so freeing it of a weight-supporting role, and also getting rid of the need for lateral undulation of the vertebral column as a component of the locomotory cycle (Carrier 1987; Kemp 2005). Basal therapsids at least facultatively had adopted the new gait, and by the cynodont grades it was obligatory.

The second change inferred from the fossil record is the evolution of the mammalian muscular diaphragm that walls the back of the pleural cavity and when brought into operation substantially increases the volume of air flowing into the lungs. Its presence is never certain, but from the therocephalian grade (Fig. 8.5) onwards the thoracic ribs form a very mammal-like ribcage, with a well-defined posterior boundary, very similar to the costal anatomy of mammals.

The third osteological correlate is the evolution of the mammalian secondary palate in the roof of the oral cavity that separates the choanal air flow above from food preparation in the oral cavity below, thereby allowing breathing to continue while feeding. The situation is complicated by other functions of the secondary palate in mammals, as it also has a role in strengthening the anterior part of the skull (Thomason and Russell 1986), and it acts as a surface for food manipulation by the tongue whilst chewing. On the mammalian lineage, the structure is first found at the cynodont grade although basal therapsids do possess a trough in the midline of the primary palate (Fig. 8.4), which probably performed the separation function to some degree.

Locomotory mechanics

The transition from a basal amniote sprawling gait to the mammalian parasagittal gait, involving a major reorganization of the musculo-skeletal arrangements (Fig. 8.2) is well illustrated by the fossil record (Kemp 1980, 1982, 2005). The vertebral column lost virtually all traces of the primitive lateral undulation characteristic of basal amniotes, a change already more or less completed within the pelycosaur grades. By the basal therapsid grade, the dorsal part of the pelvic girdle had enlarged, the hind limbs become relatively longer and more gracile, and

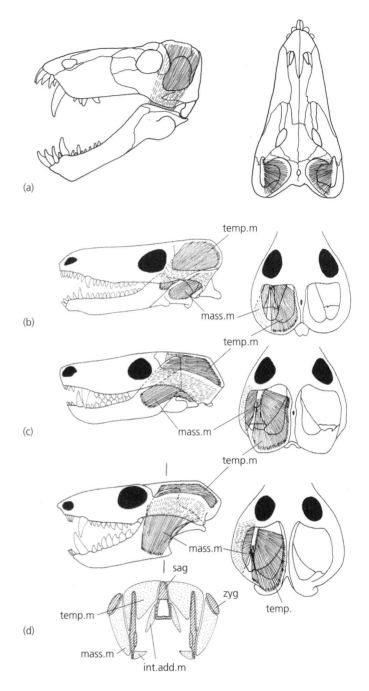

Figure 8.8 Evolution of the jaw musculature.
(a) Adductor jaw musculature of a basal therapsid.
(b) The basal cynodont *Procynosuchus*; (c)
Thrinaxodon; (d) the eucynodont *Chiniquodon*, with
coronal section through muscle mass. int.add.m =
internal pterygoideus muscle; mass.m = masseter
muscle; temp.m = temporalis muscle; sag = sagittal
crest; zyg = zygomatic arch. (From Kemp 2005.)

the feet plantigrade and capable of being placed be-
neath the body. From the limited range and amplitude of movement of the pelycosaur hind limb
evolved an adaptable system that permitted the
gait to be varied between the primitive sprawling

mode and the mammalian parasagittal mode depending on the circumstances of the moment. The
fore limb evolved less radically at first, although the
therapsid shoulder girdle had become more gracile
and loosely attached to the rib cage. This allowed a

higher proportion of the fore-limb stride to be accommodated by shoulder girdle movements, and the limb itself to remain relatively shorter than the hind limb. By modification of the shoulder joint, the range of movements of the humerus was increased compared to the strictly constrained, low amplitude fore limb stride of the more basal condition. Eventually, by the later non-mammalian cynodont grades, the fore limb did, as it were, catch up with the hind limb and was equal to the latter in gracility and variability of movement.

As inferred from the anatomy of the bones, the locomotory changes also involved very considerable reorganization of the relative sizes and dispositions of the associated muscles. Principally this involved reducing the number of muscles necessary and shifting the site of origin of the major protractor and retractor muscles above the shoulder and hip joints. The dual effect of the change was both to increase the maximum stride length and also, by the reduction of the moment of inertia of the limb, to increase the maximum stride frequency. Furthermore, their reduced moment of inertia created increased agility of the limbs and hence increased agility of locomotion. The overall effect of the evolutionary changes therefore was to produce a locomotory system which possessed greater agility plus greater speed, in keeping with the enhanced ability of mammals to operate successfully in such activities as hunting prey, avoiding predators, and living in a physically heterogeneous environment such as burrows and trees. Correlated with these mechanical changes was, of course, the enhanced stamina that came with the rise in metabolic rate associated with endothermy.

Sense organs

The transformation of the non-mammalian synapsid jaw-articulation bones, quadrate and articular, into the accessory ear ossicles of the mammalian middle ear, incus and malleus, and also the angular bone into the ectotympanic bone supporting the ear drum is the classic paradigm of the power of comparative anatomy, embryology, palaeontology, and more recently functional morphology, to explain what otherwise might have seemed a completely inexplicable evolutionary event. From the basal therapsid grade through the cynodonts and earliest mammals, several stages in the reduction in size of the quadrate and postdentary bones correlated with changes in jaw musculature and reduction of the reaction forces at the jaw hinge are documented. That this is associated with increasing sensitivity to airborne sound reception is scarcely doubted, in the light of Allin's (1975) detailed demonstration of the similar geometrical relationships of the homologous bones in non-mammalian therapsids and mammals (Fig. 8.9a–d). Subsequently, Kemp (2007) demonstrated that the sizes, masses, and detailed attachments of the reduced bones in eucynodonts are consistent with hearing relatively low-frequency but biologically significant sound via the mandible. This account of the evolution of the hearing system is consistent, indeed functionally integrated, with the simultaneous modification of the feeding function of jaws. As the jaw muscles evolved to increase the bite force and reduce the hinge reaction, so the hinge bones reduced in size and therefore increased their sensitivity to higher frequency sound.

Mammals are characterized by nasal and ethmoid turbinates within the nasal cavity which carry a large area of olfactory epithelium, and there is some evidence for their existence as cartilaginous structures in therapsids. The nasal cavity of all forms are voluminous, and bear internal ridges (Fig. 8.9e) that have been interpreted as the sites of attachment of unossified turbinates whose function would presumably have been to increase the sensitivity and range of olfactory cues detected.

Brain

It is one of the most unfortunate aspects of the origin of mammals that so little can be discovered about how the huge increase in brain size that marked the transition occurred. Compared to modern reptiles, mammals possess a brain that is anything upwards of five times larger, and that has been radically restructured by the evolution of the neocortex. The problem arises not only from the unpreservable nature of the brain and its regions, but also from the fact that in vertebrates other than birds and mammals, the brain does not anywhere near fill the endocranial cavity of the skull. Therefore not even the gross external morphology of non-mammalian synapsid brains is adequately known. Nevertheless, it is possible to gain an approximation of some of the parts,

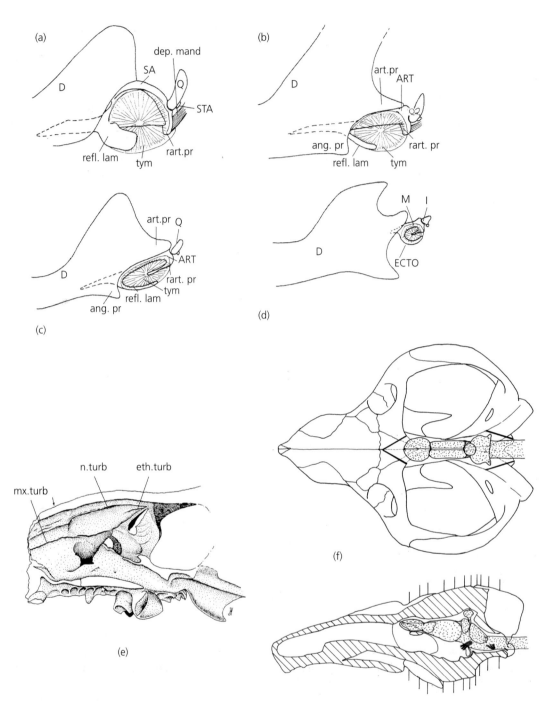

Figure 8.9 (a–d) Allin's concept of the origin of mammalian ear ossicles. (a) *Thrinaxodon* stage; (b) eucynodont stage; (c) basal mammaliamorph stage; (d) modern mammal stage. (e) Evidence of turbinates in the nasal cavity of a eucynodontian. (f) Reconstruction of the brain of the eucynodontian *Chiniquodon*. ang. pr = angular process of dentary; ART = articular; art. pr = articular process of dentary; D = dentary; dep. mand = depressor mandibuli muscle; ECTO = ectotympanic; I = incus; M = malleus; Q = quadrate; rart. pr = retroarticular process; STA = stapes; tym = tympanic membrane. ((a)–(d), from Kemp 2005, redrawn from Allin 1975; (e) from Hillenius 1994; (f) from Kemp 2009.)

and also the maximum possible size of the major regions (Hopson 1979). In pre-cynodont-grade synapsids the brain is small and tubular, lacking virtually any significant expansion of the forebrain. Even in the eucynodont-grade *Chiniquodon* (Fig. 8.9f), the cerebral hemispheres are relatively very small, no larger than expected in an amphibian, although the cerebellum was evidently approaching mammalian size (Kemp 2009). This implies that, as might be expected, the part of the brain associated primarily with fine musculo-skeletal control had evolved in correlation with the changes in jaw musculature and locomotion. In contrast, the very small forebrain volume suggests that no significant neocortex was present, and so the part of the brain most associated with complex behavioural strategies, including social behaviour, had not evolved at this stage. The enlarged forebrain had, however, evolved by the early mammaliaforms such as *Morganucodon*, implying that it is only by this point that these derived mammalian attributes had evolved, presumably in the context of small, nocturnal animals hunting insects.

Growth

There is a marked change in bone histology between the pelycosaur and therapsid grades. The pelycosaur grades have mainly lamina-zonal bone, which in modern amniotes is indicative of slow growth and is marked by lines of arrested growth presumably indicating a seasonal cessation of growth (Huttenlocker and Rega 2012). Therapsid bone is more variable, but in most cases the principal component is the well-vascularized fibro-lamellar bone characteristic of all but small mammals (Chinsamy-Turan 2012). Whether this relates directly to metabolic rates is debatable, but it does indicate that therapsids were approaching mammalian growth rates.

A high rate of juvenile provision by the parents is a necessary correlate of endothermy, and there is some evidence bearing on the origin of this behaviour. Ectothermic amniotes characteristically have indefinite growth, the animal increasing in size throughout life. In contrast, the endothermic mammals and birds have determinate growth whereby a very high rate of juvenile growth leads to a fixed adult body size, after which growth ceases. This is correlated with their respective patterns of tooth replacement. In ectotherms, shedding and replacement of teeth continue throughout life in order for the dentition to keep pace with the increasing jaw length. Modern mammals in contrast have reduced tooth replacement to a single wave, the diphyodont condition, in which the incisors, canines, and premolars are replaced once, and molars, which are not replaced, erupt later in life. Thus, there is a functional juvenile dentition during the growth phase, followed by the permanent adult dentition which serves the animal for the rest of its life. Cynodonts had indeterminate growth and continuous tooth replacement. However, in the early mammaliaform *Sinoconodon*, while the incisors and canine teeth underwent multiple replacement, the postcanine teeth were not replaced but were shed at the front and added sequentially at the back of the jaw, a condition intermediate in some respects to modern mammals. Parrington (1971) showed many years ago that a diphyletic replacement pattern was present in the early mammaliaform *Morganucodon*, with the implication that this grade had indeed evolved a high level of parental provision.

8.1.3 The pattern of evolution of mammalian characters

From the sequence of hypothetical stages developed by cladistic analysis of the synapsid fossil record, an implied pattern of acquisition of mammalian characters between the last common ancestor of reptiles and mammals and the mammals themselves can be inferred. It encapsulates all that the fossil record can reveal about the long trek through morphospace that culminated in the origin of mammals. The taxonomic resolution is not high in the sense that successive reconstructed hypothetical ancestors differ from one another morphologically to a degree typical of Linnean families or orders, and the gaps between them presumably represent long sequences of species-level transitions. Nevertheless, the resolution of the synapsid record is better than that of any other fossil sequence spanning a comparable distance through morphospace.

In a series of publications, Kemp (1982, 1985, 2005, 2007a) analysed the inferred evolutionary changes in morphology between successive nodes, noting that in every case there were significant modifications to several characters each time. These included characters clearly associated with a single definable function, but also characters associated with different functions. Most of the observable character transitions concern feeding and locomotion, which is not surprising since these mechanical functions are reflected most directly and in most detail by the skeletal anatomy but, as described earlier, there are also anatomical indications of modifications to ventilation, sense organs, brain, and mode of growth. The overall pattern of acquisition corresponds to predictions based on the correlated progression model of major phenotypic evolution: relatively small changes in numbers of characters at each stage, with no single character ever evolving without being accompanied by changes in other characters. For example, a change in dentition is invariably accompanied not only by correlated changes in the jaw musculature and articulation as would be expected, but also by changes in characters associated with other functions such as locomotion and ventilation. To take the transition between sphenacodontid pelycosaur and basal therapsid grades, evolution of several new characters of the locomotory system was accompanied by the appearance of a choanal trough, enlarged jaw musculature, and higher growth rates. Again, the transition from therocephalian grade to basal cynodont grade is marked by changes in dentition and jaw musculature, airborne sound sensitivity, pelvic girdle and hind-limb musculature, secondary palate, diaphragm, and possibly brain size.

A full picture of character evolution is constrained by the lack of evidence of the presumed numerous other intermediate stages on the lineage, and so it is not possible to show empirically whether the correlated progression pattern applied at a lower taxonomic level such as genus to genus. It is always possible that at this resolution, evolutionary modification tended to affect only one functional system at a time, for example the locomotory system or the feeding system. If so, the pattern would correspond more to a modular rather than a correlated pattern, with the system in question regarded as a semi-independent evolutionary module. The correlated changes seen at the higher taxonomic level represented in the actual fossil record would then

be the summation of an invisible sequence of independent transitions in different respective modules. However, the argument against this is that not a single transition amongst all those actually known consists of the evolution of one such putative module in the absence of accompanying changes in others. The main argument in favour of the correlated progression interpretation is an appeal to the evidence supporting the general application of this model of evolution that is derived from studies of living organisms and computer modelling, presented at length in Chapter 3. However, this conclusion is reinforced by the empirical compatibility of the synapsid fossil record with the predictions of the correlated progression model of evolution rather than with those of the modular model.

The full picture of the pattern of acquisition of mammalian characters would also include numerous characters of the soft tissues and physiology that are completely invisible in the fossils, but each of which must have evolved at some point along the lineage. The correlated progression interpretation can be expanded to embrace these as further elements in the complex network of interactions between traits that represents the integrated organism (Fig. 8.10). It is within this scheme that the empirically observable characters of the fossils can be fitted congruently.

There have been few other commentaries on the pattern of evolution of mammalian characters, and most general work has focused specifically on the origin of endothermy, with little regard for the organism as a whole (reviewed by Kemp 2006; Clarke and Pörtner 2010; Hopson 2012). Almost without exception, proposed explanations have assumed that one of the several functions of endothermy is the primary one in the sense that it evolved first, as a result of natural selection for it alone. This role is attributed by some authors to the high aerobic metabolic capacity and associated activity levels; to others selection for thermoregulation was paramount, for the purpose of invading the nocturnal environment, or developing a large brain, or increasing juvenile growth rate. The most recent proposal for a single primary function is Hopson's (2012), who suggested that the step between pelycosaur grade and therapsid grade resulted from selection for an elevated aerobic metabolic rate to allow more extensive foraging

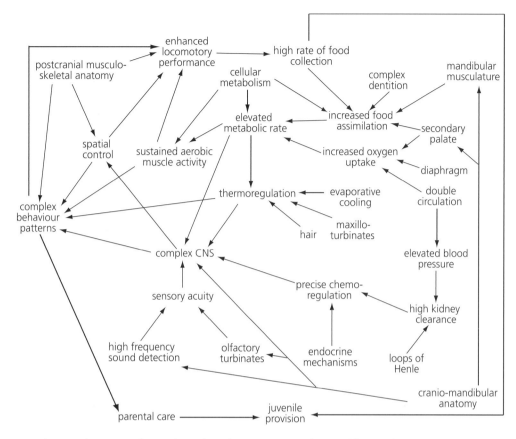

Figure 8.10 The mammal as a system: functional interrelationships amongst principal parts and functions. (From Kemp 2007.)

behaviour. Against these primary-function-type views, Kemp (2006) argued that the many traits whose interactions are responsible for endothermy, and the several functions of that physiological strategy, are highly integrated with one another (Fig. 3.6). It would therefore be impossible for any one trait or function to have evolved to a major extent in isolation of the others. Raising the metabolic rate by a small increment simultaneously increases both the maximum aerobic metabolic rate and the available heat source for maintaining a temperature gradient with the ambience in a cool environment. Simultaneously, the elevation of the metabolic rate itself above a very small increment is only possible with adjustments to mitochondria and their metabolic pathways, to the gas-exchange mechanisms from ventilation to haemoglobin biochemistry, and to the food-assimilation mechanisms from its collection through to mastication and absorbtion. It must

also be borne in mind that each of these processes is itself a system of interrelated sub-parts. Thus, even from the more limited endothermic focus of mammalian biology, the empirically observable pattern of character evolution in the synapsid fossil record of jaw function, locomotion, ventilation, brains, and sense organs fits perfectly well in the picture of the evolving scheme and so supports the correlated progression interpretation.

The question of what kind of environment and complex of selection forces was associated with the origin of the mammalian body plan was discussed in Chapter 6, section 6.4.

8.2 Birds

The world's most iconic intermediate-grade fossil unquestionably is *Archaeopteryx* (Fig. 8.15a), in part because of the fortuitous date of the discovery of the first specimen in 1861, shortly after the publication of *On the Origin of Species*, but mainly because it is indeed unmistakably intermediate in morphology. It combined what at that time were believed to be uniquely bird characters of feathers and avian wings with a series of reptilian characters, notably a long, bony tail and teeth. It also appeared to lack any specializations of its own—autapomorphies as they would now be termed—to debar it from a position of actual ancestry. Thomas Huxley (Huxley 1868, 1886) studied what by then were two specimens of *Archaeopteryx*, and to him is owed the origin of the view that *Archaeopteryx* and therefore birds as a whole descended from bipedal, carnivorous dinosaurs. For a long period, subsequent opinion was divided about avian origins, with pterosaurs, crocodiles, and basal archosaurs all proposed by one author or another (see Witmer 1991 for a review of this period). However, dating from Ostrom's (1973, 1976) now classic detailed comparison of small theropod dinosaurs and *Archaeopteryx*, the issue has since been resolved to virtually everybody's satisfaction in favour of the view that the taxon Aves not only evolved from dinosaurs but is formally a clade nested within the carnivorous, bipedal group Theropoda: birds are miniaturized, feathered, flying bipedal dinosaurs, a hypothesis first made cladistically detailed and explicit by Gauthier (1986).

The picture has expanded dramatically since establishment of the relationship between birds and theropod dinosaurs thanks in large part to the extraordinary and unexpected discovery of an array of new feathered fossils of small theropod dinosaurs, and stem birds (reviewed by Norell and Xing 2005; Xu and Norell 2006; Xu et al. 2010). The majority of these are from the Lower Cretaceous Jehol Group of western China, although they have turned up in a number of other localities including Jurassic deposits. The first such feathered dinosaur to be discovered, *Sinosauropteryx* (Fig. 8.13a), was described as recently as 1996 (Ji and Ji 1996), while by 2005 Norell and Xing (2005) in their review listed about a dozen species in at least five different theropod families. New taxa have continued to be described at regular intervals since (Xu et al. 2010). It is now beyond question that feathers, as well as numerous avian skeletal characters, evolved long prior to the appearance of crown group Aves, and that because

they predate the origin of flight, feathers can no longer be regarded either as a uniquely avian character or as initially an adaptation for flying.

Clearly much more of the avian body plan was assembled in dinosaurs than once believed. Olson's comparison of *Archaeopteryx* with the coelurosaur theropod *Deinonychus*, and with the small *Compsognathus* that occurs contemporaneously and co-geographically with *Archaeopteryx*, had already drawn attention to the early evolution of essentially bird-like hind limbs and the bipedal habit, well-developed fore limbs with reduced digits, and a number of other skeletal characters such as a relatively long, flexible neck. To these indications of bird-like terrestrial running and food capture, the prevalence of non-volant feathers in theropod dinosaurs implies that the animals were insulated. This indicates that high metabolic rates and activity levels had evolved, presumably with all the complex of structures and processes necessary for endothermic temperature physiology at least partially in place. Furthermore, evidence, if not always fully convincing, has been presented in a variety of theropod dinosaurs for a divided circulatory system, air sacs and avian-like lungs, high juvenile growth rates, and complex reproductive and social behavioural strategies. Indeed, so much of essential bird biology was established in non-avian dinosaurs that understanding the origin of birds requires consideration of the origin and radiation of theropod dinosaurs as a whole. The relationship of birds to non-avian dinosaurs is perhaps best seen as analogous to that between bats and non-chiropteran mammals: understanding the biological origin of bats requires a comprehension of mammalian evolution as a whole.

In recent years a considerable effort has been made to resolve the interrelationships of the dinosaurs and their phylogenetic position within the Archosauromorpha, with a gratifying degree of concordance (Weishampel et al. 2004; Senter 2007; Brusatte et al. 2010, 2010a; Langer et al. 2010; Xu et al. 2010; Irmis 2011; Nesbitt 2011), and Fig. 8.11 is a generally agreed cladistic framework for the most significant stem taxa of Aves. Each node represents an ancestral grade whose characters can be reconstructed on the basis of the respective subtending fossil taxon. From the series, the sequence

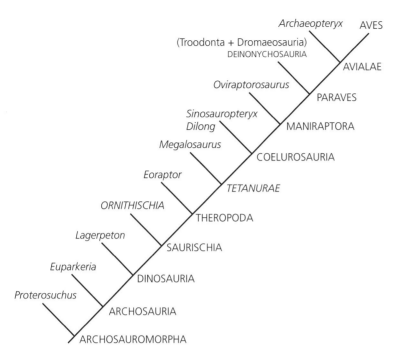

Figure 8.11 Phylogenetic relationships and major grades of stem birds.

of acquisition of avian characters can be inferred, to the extent that the fossil record is informative.

8.2.1 The grades of fossil stem birds

Basal archosaurs

The archosauromorphs as a whole constitute one of the two branches of neodiapsid amniotes, having diverged from the other, lepidosauromorphs in the Middle Permian: the earliest known archosauromorph is *Archosaurus* itself, a poorly known form from the Upper Permian of Russia. There are several well-known taxa of basal forms in the Lower Triassic such as *Prolacerta* and *Proterosuchus* (Nesbitt 2011), which retain a series of primitive features that includes a palatal dentition, a pineal foramen, and absence of the antorbital and mandibular fenestrae. Apart from some minor features of the limbs, there is nothing at this stage that clearly presages avian morphology.

By the grade represented by, for example, *Euparkeria* (Fig. 8.12a) and *Erythrosuchus*, a number of dinosaurian and avian features have appeared (Gauthier et al. 1988; Nesbitt 2011). In the skull, there are antorbital and external mandibular fenestrae,

completion of the lower temporal bar, and thecodont teeth set in sockets rather than being directly attached to the bone of the jaws. In the postcranial skeleton there are indications of the evolution of a more progressive gait, with limbs closer to parasagittal and the hind limb distinctly larger than the fore limb.

Archosaurs

The archosaurs are usually divided into the sistergroups Crurotarsi for crocodiles, phytosaurs, and a few other extinct taxa on the one hand, and Avemetatarsalia for pterosaurs and dinosaurs on the other. There are a number of early 'dinosauromorphs' which lie phylogenetically somewhere around the dinosaur stem, but whose exact relationships are uncertain (Brusatte et al. 2010a; Langer et al. 2010). *Marasuchus* (Fig. 8.12b), for example (Sereno and Arcucci 1994), is a small, carnivorous form from the Late Triassic of South America that is usually accepted as a basal dinosaur.

Dinosaurs

Characters that most unambiguously diagnose dinosaurs are postcranial, and are indicators of a

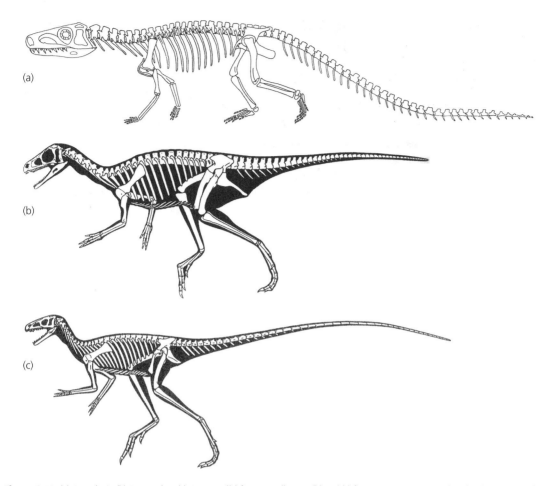

Figure 8.12 (a) *Euparkeria.* (b) *Marasuchus;* (c) *Eoraptor.* ((a) from Carroll 1988; (b) and (c) from Benton 2004 reproduced with permission from John Wiley & Sons.)

more progressive, obligatorily parasagittal gait, with bipedality at least facultatively adopted. They include the open acetabulum into which fits an in-turned femoral head, the form of the fourth trochanter of the femur for insertion of the caudi femoralis muscle, and the presence of at least three sacral vertebrae (Brusatte et al. 2010a). Also, as the taxonomic extent of feathers, or at least simple integumentary additions widens, it now seems likely that these also may diagnose dinosaurs as a whole, which of course has implications for the basic physiology of the clade. The same may well prove true of the bone histology and the indication it provides of elevated growth rates. It seems clear that even by this late Triassic stage, several of the essential elements of avian physiology and bipedal locomotion were already incipiently in place.

Saurischia

The early Upper Triassic Argentinian taxa *Herrerasaurus* and *Eoraptor* (Fig. 8.12c) are basal saurischians, and some analyses actually place them as theropods. Whichever precise relationship is true, they offer a good model for a hypothetical ancestral theropod. Both were relatively small, the first around 4 m and the second up to 2 m, and fully bipedal raptors.

Theropods

From an *Eoraptor*-like ancestor, the Mesozoic radiation of the theropod dinosaurs commenced. The great majority of taxa can be described as bipedal raptors, although occasional secondarily herbivorous and possibly scavenging forms appeared from time to time. They varied in size, details of the dentition, relatively sizes and proportions of fore and hind limbs, numbers of digits, and so on, along with large numbers of relatively minor differences in the anatomy of the cranial and postcranial skeletons. Recent detailed phylogenetic analyses have offered more or less fully resolved cladograms (Fig. 8.11) based on a large suite of skeletal characters (Rauhut 2003; Kirkland et al. 2005; Senter 2007). However, as often the case for major radiations, many of the character differences are relatively minor, homoplasy is abundant, and confidence levels in some of the phylogenetic hypotheses low. Tracing the sequence of acquisition of characters in the lineage from an ancestral theropod to the body plan of the crown birds Aves in very much detail is not yet possible, although the broad outline is growing clearer, and points on the cladogram by which some of the important avian biological characteristics had emerged are discernible.

Tetanurae

By the Middle Jurassic the Tetanurae had appeared, characterized by a tail stiffened with elongated, interlocking zygapophyses. As far as the origin of avian biology is concerned, the most important skeletal feature is the relatively large size of the hands with only three digits. The Oxford dinosaur *Megalosaurus*, the first dinosaur to be named and to come to scientific attention, is a basal member of the Tetanurae.

Coelurosauria

The majority of the Tetanurae fall into the large, diverse taxon Coelurosauria, whose members range in size from that of a small bird to the largest of the tyrannosaurids. Most were carnivores, although some lineages evolved omnivorous and herbivorous adaptations. It is of particular interest that most of the coelurosaurian taxa are now known to have possessed feathers, implying that these structures were present in the ancestral form. *Dilong* is a relatively small tyrannosurid, body length about 1.7 m, one specimen of which has simple branching filamentous feathers along its jaw and tail (Xing

et al. 2004). *Sinosauropteryx* (Fig. 8.13a), the first feathered dinosaur to be discovered, is another fairly basal coelurosaur with evidence of feathers (Ji and Ji 1996). It is also a small form, about 1 m in body length, and is closely related to *Compsognathus* that comes from the same Solnhofen locality as *Archaeopteryx*, and which was so critical to establishing the theropod roots of the ancestral bird.

Maniraptora

Maniraptorans are a clade of generally small coelurosaurs. They are well diagnosed by a series of cranial and postcranial characters. Of particular importance is the structure of the wrist, with a lunate-shaped carpal bone with a pulley surface with which the hand articulates, allowing a flexing of the fore-limb like the folding mechanism of the avian wing. They also possess uncinate processes on the ribs, and ossified sternal ribs connecting to an enlarged sterna plate, and more complex feathers.

Paraves

The maniraptoran taxon within which the final transformation to birds occurred is the group Paraves. These are small, fully feathered forms with very well-developed fore limbs. Two constituent groups are recognized, Deinonychosauria (dromasurs plus troodontids) and its sister-group Avialae which consists of the birds and whose basalmost member is agreed by most to be *Archaeopteryx*.

Deinonychosaurs were fast-running bipeds noteworthy for the long, sharp-clawed fore limbs and an enlarged, sickle-like claw on the second digit of the foot, particularly large in the dromasaurs. They were common in the Late Cretaceous, included the well-known *Deinonychus* and *Velociraptor*, and have been appreciated as close theropod relatives of birds for some time on the basis of numerous detailed skeletal similarities as well as the overall impression given by the limb proportions. A number of small deinonychosaurs with feathers preserved are known from China, such as the dromaeosaur *Sinornithornis*, which has several feather types. These include single filaments, composite bunched filaments, and ones with a central filament and serial branches (Norell and Xing 2005). The most surprising dromaeosaur discovered is *Microraptor* (Fig. 8.13b). It has multi-plumed feathers

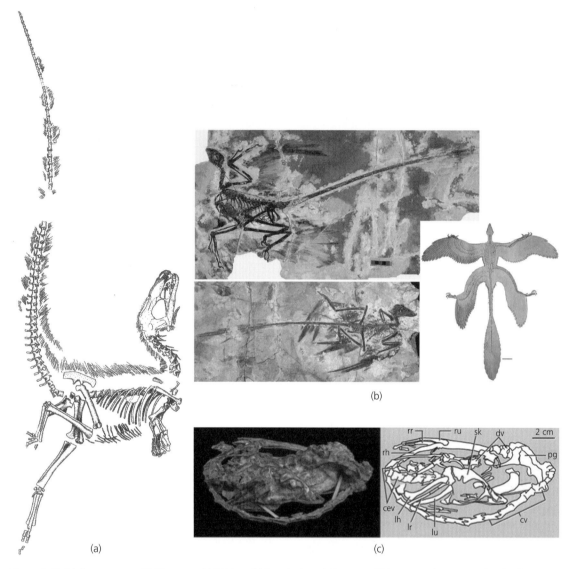

Figure 8.13 (a) *Sinosauropteryx.* (b) *Microraptor,* (c) *Mei long.* ((a) from Currie and Chen 2001; (b) from Xu et al. 2003 reproduced with permission from Nature Publishing Group; (c) from Xu and Norell 2004 reproduced with permission from Nature Publishing Group.)

around the head and body, but also asymmetrical pennaceous feathers, of the form normally associated with flight. These are found on the arms, and also the legs, as well as along the tail. It has been suggested that this 'four-winged' flying dinosaur represents a true intermediate grade in the evolution of powered flight (Chatterjee and Templin 2007). Feathered troodontids are also known, such as the small *Mei long* (Fig. 8.13c). The first fossil of

this to be discovered was found in a curled-up posture with the head tucked under the forearms in the same pose used by modern birds when sleeping, to reduce heat loss (Xu and Norell 2004). Another notable feature of troodontids, though presumably also of dromaeosaurs, is a relatively large brain that approaches that of *Archaeopteryx* in size.

Until recently, most phylogenetic analyses agreed that the dromaeosaurs and troodontids together form the clade Deinonychosauria, as assumed here so far, and that the latter clade is the sister-group of Avialae which contains effectively the birds with *Archaeopteryx* as its basalmost member. However, the picture changes regularly with the description of yet more paravian species, to the point where it is clear that a good deal of homoplasy has occurred within Paraves. In 2005, Mayr and colleagues (Mayr et al. 2005) described a new specimen of *Archaeopteryx* which is particularly well preserved, and which exhibits a number of characters of theropods in general, and one, a highly extensible second digit of the foot bearing a somewhat enlarged claw, hitherto believed to be diagnostic of dromaeosaurs. In fact, they found no characters uniquely shared by *Archaeopteryx* and an undisputed primitive avian such as *Confuciusornis* (Fig. 8.15c), and their cladistic analysis placed birds with the deinonychosaurs to the exclusion of *Archaeopteryx* (see Corfe and Butler 2006 for alternative interpretations).

In 2008, the pigeon-sized *Epidexipteryx* (Fig. 8.14a) was described from Middle to Late Jurassic deposits of Mongolia (Zhang et al. 2008). It has a shortened tail bearing four long, ribbon-like feathers presumed to be for display and, if so, this is the earliest evidence for avian-like visual display behaviour. Both shafted and non-shafted feathers are present, although none appear to have been designed for flight. *Epidexipteryx*, and the related *Epidendrosaurus* (Zhang et al. 2002) had long fore limbs close to bird-like in proportion, but with an elongated third digit on the manus which is interpreted as an arboreal adaption, as is the foot structure. These two genera, included in the family Scansoriopterygidae, combine a number of avialian features with resemblances to the more basal maniraptoran oviraptorosaurs in skull shape and enlarged anterior teeth. Zhang and colleagues (2002) interpreted these as arboreal but non-flying avialians, very close to bird ancestry and supportive of an arboreal theory of the origin of flight.

The next significant discovery was that of *Anchiornis* (Xu et al. 2009), the smallest known paravian at an estimated 110 grams body weight (Fig. 8.14b). The proportions of its limbs are similar to those of basal birds, and it is remarkable for having evolved a wrist that is incipiently bird-like, with a large radiale and a semilunate carpal attached to

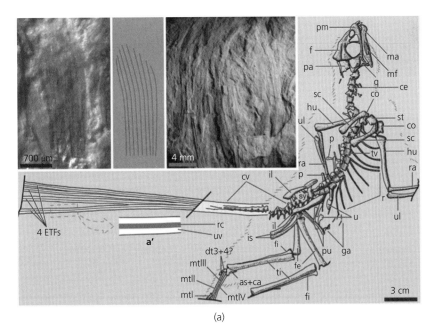

(a)

Figure 8.14 (a) *Epidexipteryx*. (b) *Anchiornis*; (c) *Xiaotingia*. ((a) from Zhang et al. 2008; (b) from Hu et al. 2009; (c) from Xu and Norell 2004, all reproduced with permission from Nature Publishing Group.)

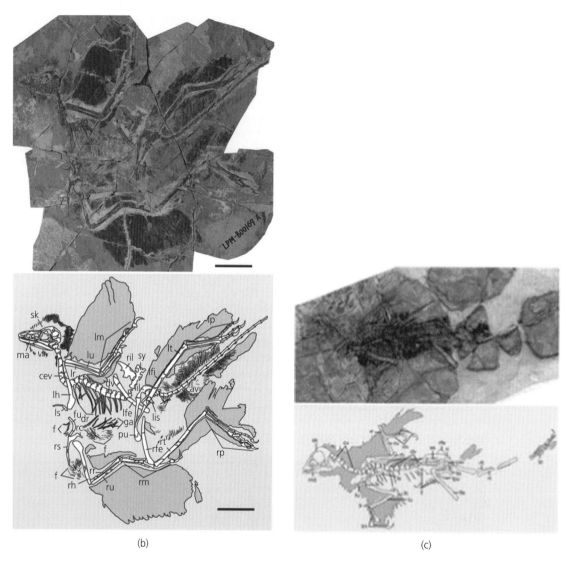

(b)   (c)

Figure 8.14 (*continued*)

the metacarpals. This suggests that the equivalent of wing-folding was possible. It has large pennaceous feathers on forearm, and on the lower leg as in *Microraptor*. Against these avian features, *Anchiornis* confusingly possessed a somewhat troodontid-like tail and was in fact first identified as a member of the Troodontidae, despite the presence of a number of more general paravian characters.

In 2011, Xu and colleagues (2011) described *Xiaotingia* (Fig. 8.14c), a very *Archaeopteryx*-like Late Jurassic theropod. The particular combination of characters found in *Xiaotingia* has proved to have a radical effect on the phylogenetic analysis of the maniraptorans and in particular on the relationships of *Archaeopteryx* itself. There are certain characters shared with *Anchiornis* and *Archaeopteryx*, and others with these genera plus deinonychosaurids. Further features are otherwise found in various combinations of basal aviales, troodontids and oviraptors. The cladogram presented by Xu and colleagues (2011) groups *Archaeopteryx*, *Xiaotingia*, and *Anchiornis* together as the sister-group of

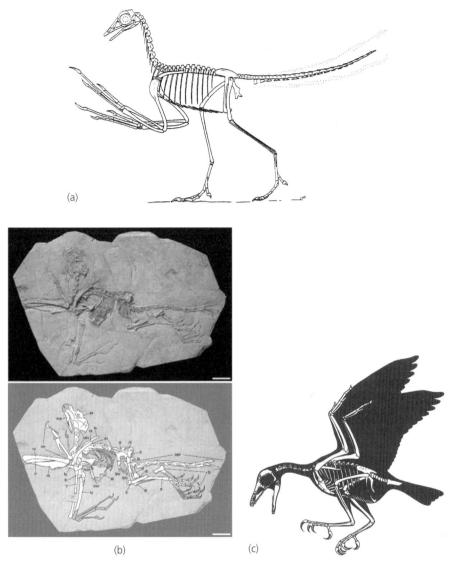

(a)

(b)

(c)

Figure 8.15 (a) *Archaeopteryx* reconstruction (from Ostrom 1979 reproduced with permission from John Wiley & Sons). (b) *Jeholornis* (from Lefevre et al. 2014 reproduced with permission from John Wiley & Sons). (c) *Sinornis* (from Sereno, P. and Rao, C. 1992 reproduced with the permission of the American Association for the Advancement of Science).

the deinonychosaurs, and these two constitute the sister-group of an Avialae in which the Scansoriopterygidae is now the basalmost member. This relationship bears the implication that active flight either evolved more than once or was secondarily lost in certain lineages. However, the level of statistical support for the cladogram is low, indicating that there is a great deal of homoplasy involved,

and a subsequent analysis using different methods has reaffirmed the avialan rather than deinonychosaurian relationship of *Archaeopteryx* (Lee and Worthy 2011). The ambiguity of relationships amongst these forms is not surprising, since all the characters are skeletal, and in most cases concern relatively small differences; for example, the relative lengths of digits and phalanges, the sizes and shapes of cranial openings, and the degree of development of various crests and ridges on limb girdles. Perhaps, however, this is not even unexpected. Clearly the ancestral paravian, the common ancestor of both avialians and deinonychosaurians, was small, long-armed, and fully feathered, probably including feathers with rachis and barbules suitable for airborne locomotion. Maybe gliding was common, and evolution of more controlled powered flight required less final evolutionary modifications than usually supposed and so arose convergently in several lineages.

Avialae

The systematics of the basalmost birds is as confusing as that concerning their immediate paravian forerunners, as more and more Late Jurassic and Cretaceous forms are described, and the possible identification of *Archaeopteryx* as a deinonychosaur exacerbates the situation. The Early Cretaceous *Jeholornis* (Fig. 8.15b) was a relatively large bird, approaching 1 m including the tail, which was even longer than that of *Archaeopteryx*, and stiffened by elongated prezygapophyses and chevrons as in deinonychosaurs (Zhou and Zhang 2002, 2003). However, it was more progressive than *Archaeopteryx* in several characters associated with power flying ability. The fore limb is relatively longer than the hind limb, the hand shorter and more robust, the coracoid a bird-like strut shape, and the carpo-metacarpus fully fused. In contrast, the hind limb of *Jeholornis* is quite similar to that of *Archaeopteryx*. Interestingly, seeds are preserved in the body cavity, indicating an herbivorous or omnivorous nature. This corresponds with a very reduced dentition and robust mandibles, and certainly suggests the presence of an avian muscular crop. As in *Archaeopteryx*, there are pneumatic spaces in the skull and vertebrae, and in this case in the sternum, suggesting the possibility of an avian interclavicular air sac.

All cladistic analyses place *Jeholornis* as a basal avian although it still lacks a number of modern avian characters. The crow-sized *Confuciusornis* (Fig. 8.15c) is one of the best known of all the Early Cretaceous birds. Derived characters include a horny beak in place of teeth, caudal vertebrae reduced to a pygostyle, and a well-developed fully perching foot. By this stage the basic avian body plan has been more or less fulfilled.

8.2.2 The origin of the avian body plan

The bird body plan is a composite of numerous characteristics well-integrated with one another. Although most attention is usually paid to the particular attribute of powered flight, this characteristic is totally dependent on other attributes. Endothermic physiology with its capacity to deliver the high power requirements is obvious, but no less of a *sine qua non* for the avian mode of powered flight are bipedality in order to free the fore limbs for flight, small body size, reduced skull weight, a feeding mechanism able to deliver a high rate of food intake, and a large brain for control of movement in the air. More indirectly, it is necessary to have parental incubation of eggs, and care and provision for juveniles. Arguably also, the habit of arboreality may have been a requirement for the evolution of flight, at least in the later stages, if not initially. No adequate account of the origin of birds can omit taking into account the evolution of all these attributes and the way in which they functionally interact.

Endothermy

Powered flight is completely dependent upon an endothermic temperature physiology capable of delivering the very high level of aerobic metabolism necessary to sustain a sufficiently high power output by the muscles. As in mammals, endothermy also ensures the maintenance of the constant, elevated body temperature by internal thermoregulation that is necessary for the operation of a relatively hugely enlarged brain and precise neuromuscular flight control mechanisms. Indeed, so fundamental is endothermy to avian life that it is inconceivable that it did not pre-date, or at least evolve in synchrony with powered flight. To achieve full endothermy, a surprisingly large number of evolutionary changes in structures and processes, from the molecular to the gross morphological, had to occur, as discussed in the context of the evolution of endothermy in mammals (section 8.1.2). At the cellular level were profound modifications to metabolic pathways, involving many enzymes and their activity profiles. At the whole organism level are the multi-organ-based physiological regulatory mechanisms that allow more or less instantaneous adjustments to rates of heat loss and heat conservation needed to deal with the range of metabolic rates occurring at different moments in the organism's life. At one extreme is the resting metabolic rate while the animal is inactive; at the other is an up to 20-fold increase to the maximum rate during active, flapping flight in a warm ambience. This temperature regulation involves a variety of internal sensory, endocrinal, and central nervous mechanisms associated with variations in blood flow, degree of filo-erection, posture, and behavioural responses. Then there is the requirement for the very high average daily rates of food assimilation and oxygen uptake, both of which involve the integrated activity of another range of structures and processes, including the necessary sensorimotor system to activate and coordinate it all. A final special requirement of the avian body plan is a reproductive system that includes a high level of parental investment while the juvenile is developing the whole complex system for itself.

Most of the evidence of temperature physiology in dinosaurs is ambiguous and there is a continuing argument about whether they were ectothermic with low, reptilian metabolic rates, or endothermic with high avian rates, or possibly some intermediate strategy. For example, in the same, relatively recently edited volume, Chinsamy and Hillenius (2004) argued for essentially ectothermic dinosaurs while Padian and Horner (2004) were persuaded that they were endothermic. Anatomical features of non-avian dinosaurs variously claimed to indicate endothermy include insulating feathers, and possible structures associated with avian-style lungs, ventilation mechanism, heart structure, and circulatory system. More indirectly, arguments have involved estimates of locomotory energetics and brain size. Finally, there is evidence from bone histology that leads to inferences about growth rates, and the possible relationship of these to temperature physiology.

The basalmost theropod dinosaur known to have possessed feather-like integumentary structures is the Early Cretaceous tyrannosaurid *Dilong* (Xing et al. 2004). Filamentous, simply branching structures are preserved along the tail and lower jaw

and it is difficult to interpret them as anything other than a layer of feathers of primitive structure. Since the fore limbs are relatively small and the adult body length about 2 m, there is no possibility that *Dilong* could have glided, let along undertaken powered flight, and in any case feathers of this structure would not have performed aerodynamically in the way that avian flight feathers do. Assuming most of the animal's body was covered with these simple feathers, then whatever additional behavioural functions they may have had, they could hardly have failed to have provided an insulatory layer, and therefore been involved in heat retention to a significant degree. Feathers have now been described in members of all the principal coelurosaurian taxa, including compsognathids, oviraptorosaurs, troodontids, and dromaeosaurs (Long and Schouten 2008), and so it is certain that feathers were a diagnostic character of at least Coelurosauria, and therefore that they evolved long before flight.

Both the basal and the maximum aerobic metabolic rates of modern endotherms is typically between five and ten times that of similar-sized ectotherms, and this requires correspondingly higher ventilation rates. The avian lung is a compact, non-compliant structure that operates via a unidirectional through-flow of air pumped between large, highly compliant air sacs. Several lines of evidence suggest that non-avian theropods possessed a bird-like system. First, the presence of pneumatic foramina opening into air spaces in vertebrae and ribs have long been known in theropod dinosaurs, and they suggest the presence of associated air sacs. O'Connor and Claessens (2005) showed a pattern of pneumaticity in the basal theropod *Majungatholus* very similar to that in modern birds, a pattern associated in the latter with diverticula of both cervical and abdominal air sacs. In fact this same pattern seems to be present throughout the non-avian theropods. Evidence for the presence of an avian-like lung comes from the nature of the thoracic ribs which appear to have restricted any expansion of the thoracic cavity, implying that the lung mechanism was avian rather than crocodilian in form and function (Schachner et al. 2009). The presence of uncinate processes on the thoracic ribs (Codd et al. 2008), which in birds form part of the pumping mechanism, plus the arrangement of the articulations of the thoracic ribs to the vertebrae (O'Connor and Claessens 2005) indicate that even in the absence of a bird-like keel on the sternum, a functional pump moving air between the air sacs was probably present. On the other hand, several researchers are not impressed by this evidence since clearly the fully developed avian mechanism of ventilation had not evolved at this stage. Ruben and colleagues (1997, 1999), for example, argued on the basis of what they interpret as soft-tissue evidence of the position of the liver in a specimen of the compsognathid *Scipionyx* that non-avian dinosaurs had expandable lungs without pulmonary air sacs, and a crocodile-like diaphragm activated by movements of the liver behind the thoracic cavity. There is also the failure to discover preserved maxillo-turbinates inside the snouts of dinosaurs, structures that are associated with water conservation in modern endotherms, both birds and mammals (reviewed in Chinsamy and Hillenius 2004).

Fisher and colleagues (2000) claimed to have identified by computer tomography (CT) scanning a three-dimensionally preserved heart of the basal ornithopod dinosaur *Thescelosaurus*, and that it showed a completely four-chambered condition along with a single aorta. If correct, it appears to confirm yet another avian feature associated with the high metabolic rates of endothermic physiology. Sadly, subsequent application of a series of modern techniques including further CT scanning and geochemical analysis has demonstrated that the structure is almost certainly a non-organic concretion of no circulatory significance (Cleland et al. 2011). However, it has been argued that the height of large sauropods is such that when the neck was held upright it would have been impossible to pump blood to the head without the magnitude of blood pressure available only in a complete double-circulatory system. Whether this argument can reasonably be extended to non-avian theropods is, of course, open to question.

There is an old, informal argument that the locomotion of dinosaurs involving a parasagittal gait is a high-energy mode of locomotion that would require a high level of aerobic metabolism, and therefore endothermic physiology (Bakker 1972). With the use of a model that links locomotory anatomy to the metabolic costs of walking and running, along with estimates of muscle volumes, Pontzer and colleagues (2009) calculated the metabolic cost of sustained bipedal locomotion at different speeds in a number of non-avian dinosaurs. They showed that locomotion at a moderate running speed required an aerobic metabolic rate typical of endotherms for all body sizes except the smallest they tested, which was *Archaeopterx*. In the case of large species, even walking required an endothermic metabolic rate, although the smaller ones could walk at a metabolic rate equal to the maximum level found in modern ectotherms. These results more rigorously support the view that non-avian bipedal locomotion did indeed require the energy output available only to an endothermic animal.

When it was first established that dinosaurs generally had the kind of bone histology associated with mammals and large birds, namely well-vascularized fibro-lamellar bone, it was taken as evidence for endothermy. However, subsequent studies threw this simple conclusion into doubt because actually amniote bone histology is very variable. Dinosaur-like fibro-lamellar bone is often found in parts of ectotherms such as crocodiles, while conversely, the supposedly ectotherm-related lamellar-zonal histology occurs extensively in small birds and mammals. The current picture is that the bone histology of a particular bone or stage of growth gives a good indication about the rate of bone growth, but that the relationship between rate of growth and the animal's metabolic rate is unclear. To some, such as Chinsamy and Hillenius (2004), the evidence of bone histology is too ambiguous to be helpful in deciding on metabolic status. To others, such as Erickson and colleagues (2009), who compared the bone histology of a range of maniraptorans and basal birds, including *Archaeopteryx*, the evidence points to a level of endothermic metabolic rates at the lower end of the range found in modern endotherms, for example in kiwis and marsupials.

Bipedality

The transformation from the quadrupal gait of ancestral archosaurs to the specialized version of bipedality found in the birds is illustrated by several grades of non-avian theropods. In a basal archosaur such as *Euparkeria* (Fig. 8.12a), the hind limb is significantly longer than the fore limb and probably moved in a parasagittal plane at least facultatively, as in the erect gait of modern crocodiles. The mode of action is described as hip-driven because most of the stride is due to the femur rotating in the acetabulum. Bird bipedality in contrast is described as knee-driven because the femur is much shortened and held close to horizontal while most of the stride is due to flexion and extension of the knee joint. This shift is presumed to be correlated with a shift in the position of the animal's centre of gravity. In the course of the lineage leading to birds, the tail reduced in size and the fore limb increased, both of which tended to shift the centre of mass forwards. To compensate, it was necessary for the position of the foot also to move forwards, which was accomplished by the increasingly horizontal orientation of the femur. At the same time, the increase in the relative length of the distal parts of the hindleg reduces the moment of inertia of the limb and hence increases its capacity for more rapid strides. Other osteological characters of the bird hind limb are the increased number of sacral vertebrae, the rotation of the pelvis so the pubis comes to project caudally, and the elongated metatarsals. There was a corresponding modification of the hind-limb musculature, as the retractor muscles shifted from the caudal vertebrae to the pelvic girdle. All these avian modifications can be recognized in various stages within non-avian theropods. By the maniraptoran grade, the tibia is longer than the femur and the metatarsals more than half the femoral length. The pubis parallels the ischium in its postero-ventral orientation and there are six sacral vertebrae. By *Archaeopteryx* (Fig. 8.15a), the relative hind-limb proportions have achieved more or less avian levels, although there are still some non-avian features such as the presence of an ascending process of the astragalus (Mayr et al. 2005). There is actually a good deal of variation and apparent homoplasy in theropod hind-limb morphology, including overlap between non-avians and avians, which led Hutchinson and Allen (2009) to conclude that the evolution from a basal theropod condition to the fully bird-like limb occurred by a series of gradual transitions rather than a single or small number of more radical shifts.

Small body size

There is no doubt that birds are miniaturized theropod dinosaurs: the body weights of the basal avialan *Archaeopteryx*, and the basal avian *Jeholornis* are estimated to have been around 700 g. It is equally certain that small body size is a prerequisite for powered flight because of the relationship between body mass (rising as the animals's linear dimensions cubed) and the power required to generate lift using muscles (rising as the linear dimension raised to the fourth power). The question is therefore whether the reduction in body size occurred prior to the evolution of powered flight or in association with it. A majority of dinosaur lineages experienced size increase during the Mesozoic (Carrano 2006), but the most marked exceptions are within the Theropoda. The ancestral coelurosaur is inferred from cladistic analysis to have been relatively small; for example, the basal tyrannosauroid *Dilong* was only about 1.6 m in body length and could not have weighed more than 10–15 kilograms (kg), notwithstanding the 6000 kg of the largest members of the family (Sereno et al. 2009). When estimated body sizes are plotted onto a cladogram of paravians, the most parsimonious weight for the ancestor is around 700 g and 70 cm in length (Turner et al. 2007). Although several derived lineages of both dromaeosaurs and troodonts evolved secondarily increased body size from this point, there was a continued general reduction within the paravian clade in several lineages independently (Lee et al. 2014; Puttick et al. 2014). It seems therefore that the avialians and within them the birds retained miniaturization from a grade that evolved prior to the origin of powered flight, and was therefore an adaptation for a quite separate ecological purpose. Benton (2014) proposed that this purpose was arboreality, and in the context of this new habitat, several lineages of paravians evolved various degrees and modes of airborne ability, including the avialan lineage which adopted powered flight.

Reduced skull weight and feeding

Adaptation for feeding in modern birds is focused on reduction of weight, in keeping with the requirements for flight, yet several of the particular morphological features appeared in the avian lineage long before flight. The antorbital fenestra in front of the orbits and the mandibular fenestra of the lower jaw are characters of archosaurs, while the second fenestra in the snout region, the maxillary fenestra, appeared at the tetanuran stage. It is still discrete in *Archaeopteryx*, although it has fused with the external nostril in later birds. Reduction in the size and number of teeth occurred frequently within maniraptorans, and the long, flexible neck that is associated with avian food collection is also characteristic of this grade. There is no direct evidence for the presence of a muscular crop in non-avian dinosaurs, although Else (2013) speculated on the possibility that such a structure or its equivalent must have been necessary for juvenile provision.

Enlarged brain

A correlation between brain size and endothermic temperature physiology is evident from the fact that birds and mammals both have a brain some ten times the volume of similar-sized modern reptiles. Brain volume as a whole and details of the sizes of the different parts are notoriously difficult to measure in non-avian. However, with the advent of CT scanning techniques, more information is becoming available. The reconstructed brain of *Archaeopteryx*, based on the endocast of the London specimen, has an estimated total volume at the lower end of the range of modern birds, but around three times that of equal-sized reptiles (Alonso et al. 2004). The cerebral hemispheres are enlarged over reptiles although less so than in modern birds. The optic lobes are relatively very large, displaced laterally, but unlike in birds they are not rotated posteriorly, while the cerebellum is also enlarged. Perhaps not unexpectedly then, the brain of *Archaeopteryx* can be described as primitive avian, and presumably was adequate for the degree of flight control achieved by this stage. The brain of the maniraptoran oviraptorosaur *Conchoraptor* has also been reconstructed (Kundrat 2007). In relative size it appears to be actually larger than that of *Archaeopteryx*, and therefore lies comfortably within the size range of modern birds. The cerebral hemispheres are enlarged to about the same proportionate extent as *Archaeopteryx* and are significantly greater than the estimates for non-maniraptorans. The cerebellum is also enlarged and there are signs that foliations were beginning to develop over its surface. A number of

other, less well-preserved endocranial casts indicate that generally speaking, brain size and degree of development of the cerebral hemispheres and cerebellums of non-avian theropods had evolved significantly towards the avian condition.

Incubation and parental care

The preservation of a number of oviraptorids and troodontids in association with nests of eggs has given some insight into parental behaviour in maniraptoran theropods, which appears to have been remarkably bird-like (Chiappe and Dyke 2007). Specimens have been found apparently sitting on the clutch in an avian fashion, legs tucked under the body in the centre of the clutch and the fore limbs surrounding them. This is certainly very suggestive of incubation. Varricchio and colleagues (2008) noted that the clutch size of *Troodon* and oviraptorids is 20–30 eggs which is the range found in modern species such as the ostrich where the male provides the parental care. Furthermore, the incubating specimens lack histological signs of bone absorbtion of the kind expected in a female that has recently produced a large clutch of well-calcified eggs. Other avian features include the asymmetric shape of the eggs and their multi-layered shell, and evidence for serial egg-laying.

This evidence about reproductive habits certainly suggests that avian-type behaviour evolved much earlier than birds and flight, and that they were associated with a high incubation temperature.

Arboreality

There is little doubt that the ancestral bird condition included at least facultative arboreality, but that non-avian maniraptorans did not. *Jeholornis*, for example, possessed a reversed hallux, and long claws on the hind digits (Zhou and Zhang 2002) as did *Confuciusornis* (Zhang and Farlow 2001), and a comprehensive morphological analysis showed that these basal birds cluster in morphospace with modern perching, ground-foraging birds rather than with fully ground-living forms (Dececchi and Larsson 2011). On the other hand, surprisingly perhaps, *Archaeopteryx* clusters along with other non-avian maniraptorans with modern ground-based birds and mammals, indicating on this evidence that it was not arboreal, a view long ago proposed by

Ostrom (1976). Currently the most basal non-avian with arboreal adaptations is the scansoriopterygid *Epidendrosaurus* (Zhang et al. 2002). Whether this was a convergence with birds, or whether *Epidendrosaurus* is more closely related to birds than is *Archaeopteryx*, as Xu and colleagues (2011) argued, is not clear. In either case, it implies that arboreality evolved subsequently to the *Archaeopteryx* grade and that flight at least to the level of *Archaeopteryx*'s ability did not require the help of trees to gain initial height.

Flight

Despite the enormous amount of attention that has been and continues to be paid to the origin of bird flight, there is still disagreement about which scenario represents the true sequence of acquisition of full, avian-style powered flight. In part this is due to the artificiality of firm distinctions between parachuting, passive gliding, and active powered flight modes of locomotion (Hutchinson and Allen 2009). Traditionally the argument was between two clear-cut alternatives. Proponents of the 'trees down' scenario speculated that small, arboreal non-avian dinosaurs first evolved a gliding wing for moving from tree to tree and descending, and then gradually elaborated it into the powered flight structure supposedly found in *Archaeopteryx*. The alternative was the 'ground up' hypothesis that envisaged a small, cursorial, non-avian theropod increasing its locomotory ability by evolving incipient wings to carry it over obstructions. A third, more recent model is described as wing-assisted incline running (Dial 2003; Dial et al. 2006). This is derived from observations of hatchlings of certain ground-living birds, which can generate enough aerodynamic forces by flapping their incompletely developed wings to increase traction while running up steep slopes, even to the extent of climbing up tree trunks. For each of these proposals, there is anatomical evidence and aerodynamic arguments both in favour and against. Furthermore, all of them experience the problem of explaining what the function of the initial, presumably incipient stages of evolution of the wings was before they were adequate for any of these proposed pre-flight stages.

To consider the question from first principles, fully evolved avian powered flight requires a complex of attributes (Padian and de Ricqlès 2009). These include a high enough power output from the muscles, a hind limb capable of supporting the body before take-off, and a small body size, all of which were evidently present in the lineage prior to origin of flight. Also, there is a requirement for a rapidly reacting sensori-motor system for 3-D orientation and maintenance of stability, which must also have already evolved to some degree in bipedal, non-volant runners. The principal remaining necessity was enlargement of the fore limb and the evolution of specifically flight feathers with asymmetric vanes and barbules to form the aerofoil surface, plus modification of the bones, joints, and musculature of the forearm and shoulder girdle to generate the wing beat.

The uncertainty about the interrelationships amongst paravians, and for that matter the maniraptorans as a whole, has prevented a simple, clear description of the sequence of acquisition of these directly flight-related characters, a situation compounded by the question of what role if any a four-winged stage might have played in the evolution of the fully avian flying mechanism.

Cladistically, *Microraptor* (Fig. 8.13b) is a dromaeosaur. Unlike other members of the group, it possessed fore limbs as long as the hind limbs comparable to the proportions of *Archaeopteryx*. All four limbs carried veined, barbed, asymmetrical feathers similar to the primary and secondary flight feathers of modern birds. The effect of the asymmetry of the vane is to reduce the drag on each feather, making the wing as a whole more efficient at generating lift. There are also asymmetrical feathers on the tail. Tests by computer simulation (Chatterjee and Templin 2007), physical models (Alexander et al. 2010), and aerodynamic calculations (Palmer 2014) have demonstrated the feasibility of gliding by means of the four wings, whether the hind wings are held below and a little behind the fore wings like a bi-winged aeroplane, or are positioned completely behind the fore wings in a tandem manner. The models achieved an average glide angle of about 14° and a speed of around 6 m.s^{-1}, which are within the range of gliding abilities of modern birds.

Anchiornis (Fig. 8.14b) also has a fore limb comparable to that of *Archaeopteryx* and it too had feathered hind limbs (Hu et al. 2009). However, their feathers were not so long or extensively distributed as in

Microraptor, being restricted to the tibia and metatarsals, plus small pedal feathers. Nor did they have the asymmetrical feathers.

The hind leg of *Archaeopteryx* also has asymmetrical, vaned, aerodynamically designed feathers, but they are restricted to the femur and tibia of the more proximal part of the leg. From the anatomy of the pelvic region, Longrich (2006) argued that the hind legs of *Archaeopteryx* could be abducted sufficiently for their flight feathers to contribute as much as 12 per cent of the total aerofoil area, and therefore to have had a significant effect on decreasing stalling speed and turning circle.

The sequence *Microraptor–Anchiornis–Archaeopteryx–*birds could therefore be taken as a morphosequence of the evolution of fully avian flight from a four-winged gliding ancestry. However, *Anchiornis*'s actual phylogenetic position is unclear, for it possesses a number of characters, particularly of the skull and dentition, otherwise found in the non-flying troodontids (Hu et al. 2009; Lee and Worthy 2011).

The two genera included in the Scansoriopterygidae (Fig. 8.14a), the sparrow-sized juvenile *Epidendrosaurus* (Zhang et al. 2002), and the pigeon-sized *Epidexipteryx* (Zhang et al. 2008), confuses even further the question of the origin of avian flight. Cladistic analyses indicates that scansoriopterygids are avialans, either at the base as the sister-group of *Archaeopteryx* plus avians (Zhang et al. 2008; Hu et al. 2009; Xu et al. 2011) or nested within birds (Lee and Worthy 2011). They have a small sternum like that of the basal bird *Jeholornis* (Fig. 8.15b), and the fore limb is as relatively long as that of *Archaeopteryx*, with the humerus at least as long as the femur. Otherwise, however, there is no evidence of any flight capability. No shafted feathers, symmetrical or asymmetrical, are preserved on fore or hind limbs even though, at least in the case of *Epidexipteryx*, non-vaned filamentous feathers are present over the surface. Four extraordinary elongated vaned feathers are carried by the tail, presumably for behavioural purposes. The hand of *Epidendrosaurus* is long and well-clawed, which suggests that it is adapted for climbing (Zhang et al. 2002). On the other hand, though not necessarily contradictory to a climbing ability, the claws on the hind feet of *Epidexipteryx* fall within the morphological range of

modern ground-dwelling birds (Zhang et al. 2008). Possibly scansoriopterygids secondarily lost the power of flight, or possibly *Archaeopteryx* evolved it independently of birds.

At any event, *Archaeopteryx* is still the best candidate for the most primitive bird-like flier yet known; even if its hind limbs were an integral part of its flight mechanics, the fore limb certainly dominated. The consensus is that the size of its wing and the presence of asymmetrical feathers in a pattern resembling modern flight feathers points to quite good flying ability including, at least to some extent, active flapping flight. *Archaeopteryx* also demonstrates an intermediate condition in the transition of the musculo-skeletal system of the shoulder and fore limb from one presumably used for predatory prey capture in non-flying maniraptorans, to one capable of active flapping and fine aerodynamic control. The fully expressed avian condition required the evolution of a keel on the ossified sternum to accommodate a large pectoralis muscle for the downstroke of the wing beat. Also, the orientation of the saddle-shaped glenoid socket of the shoulder joint changed so that the fore-limb movement relative to the bird's body axis is up and down, rather than to and fro as in tetrapods. A third development was the evolution of an acrocoracoid process on the shoulder girdle, a substantial anterior process of the coracoid (Fig. 8.16). This plays a highly important role in modern bird flight (Baier et al. 2007). It is the point of attachment of a strong ligament, the acrocoraco-humeral ligament, which stabilizes the head of the humerus in the glenoid cavity during the downstroke. It also provides an origin for the coracoid head of the biceps muscle, which becomes positioned anterior to the fore limb and so its contraction causes the wing area to be reduced without a ventral bending of the limb that would lead to loss of aerodynamic efficiency, and is also essential for the wing-folding mechanism during rest. The acrocoracoid also creates the triosseous foramen, through which the tendon from the supracoracoideus muscle passes, thereby converting this muscle to an active wing elevator. In maniraptorans generally, the shoulder girdle retained the condition of bipedal theropods (Fig. 8.16a–c). In contrast, the wing mechanism in *Archaeopteryx* was intermediate in several respects between basal maniraptorans

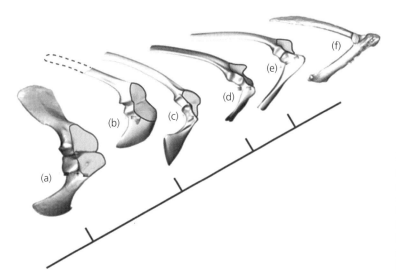

Figure 8.16 Evolution of the avian scapula-coracoid. (a) Alligator. (b) *Sinornithoides*. (c) *Sinornithosaurus*. (d) *Archaeopteryx*. (e) *Confuciusornis*. (f) Pigeon. Note the reduction of the bone for muscle attachment anterior to the glenoid, and its replacement by the development of the anteriorly projecting acrocoracoid process of the coracoid. (From Baier et al. 2007.)

and avians (Fig. 8.16d). Although at this stage there was no keel developed on the sternum, the scapula and the coracoids were both slender and bird-like, and orientation of the glenoid is intermediate between longitudinal as in tetrapods and vertical as in birds. Finally, there is a small but distinct acrocoracoid, although it still lies somewhat below instead of in front of the glenoid.

8.2.3 The pattern of acquisition of avian characters

The most striking message from the fossil record of the ancestry of birds is that most of the essential features of the avian body plan and biology evolved prior to the origin of flight, even of gliding flight. For as long as small body size, feathers, and long forearms were unknown in any archosaur prior to *Archaeopteryx*, which was of course assumed to have been a flying animal, the view was that the emergence of birds could only be understood in the light of adaptations for flight. It must now be accepted that bird biology evolved originally for a mode of life other than flying, and that active flight is just, as it were, one of the adaptive options available to the deinonychosaurian, even perhaps the maniraptoran body plan. The picture that is emerging is one of an evolutionary pattern of character acquisition broadly comparable to that seen in the origin of mammals 50 million years earlier. This, as discussed earlier,

consisted of the evolution of numerous correlated character states associated with an endothermic metabolic rate and the complexity, elevated activity level, internal homeostasis, high rates of embryonic and juvenile development, and adaptable behaviour that endothermy facilitated. Anatomical evidence from the fossil theropod lineage indicates evolutionary changes related to increased rates of food intake and ventilation. The adoption and refinement of bipedality led to more agile and potentially faster locomotion, comparable to the evolution of the parasagittal gait in mammals, and of benefit for food capture and predator avoidance. Equally telling is the evolution of an enlarged brain, again as in mammals, and again an integrated part of the organism. The brain simultaneously required a high metabolic rate and thermoregulation, contributed to the high metabolic rate, and was a necessity for the agile locomotion and behavioural repertoire manifestly present within maniraptoran and earlier grade dinosaurs.

Within the paravian-grade radiation, small, insulated, highly active bipeds, with complex parental, social, and community behavioural patterns had evolved. At least some of these used their agility and lightness for an arboreal mode of life, and various modes of passive parachuting and gliding were evident. From this stage, the transition to powered flight required another series of correlated evolutionary changes. Further enlarged fore limbs with complete flight surfaces using specially modified feathers, more precise aeronautical adjustments of angle of attack and exposed wing area, and three-dimensional sensorimotor control mechanisms coupled with precise central nervous regulation must in principle have evolved in a correlated progression. Some aspects of these modifications are preserved in the fossil record, such as the acquisition of the avian shoulder mechanism (Baier et al. 2007), and the shift via the long stabilizing, feathered tail of *Archaeopteryx* to the modern avian fanned tail (Clarke et al. 2006).

In the terms of the alternative models of character acquisition, at a superficial level the origin of birds might be viewed as in part modular, because bipedalism for example evolved prior to flight, and therefore the hind and fore limbs may be regarded as semi-independent modules. However, this oversimplifies the picture. Attributes that evolved along with the limb mechanics of bipedal locomotion include many that are also necessary correlates of flight. Elevated metabolic rates and all that this implies from cell structure to feeding and ventilatory functional morphology, along with neurosensory control of an unstable locomotory mode, have been discussed. Therefore, when flight did evolve, it is meaningless to regard the fore limb as a semi-independent module because from the very initiation of its new mode of action it was deeply integrated with the rest of the organism. Furthermore, as it evolved it necessarily demanded new correlated modifications elsewhere in the organism. To mention just a couple of examples, the shift in orientation of the body axis from vertical to horizontal moved the centre of mass which the hind-limb pose had to accommodate, and mechanisms of heat loss had to cope with the higher energetics and heat output of active flight. Much more realistically, the transition from basal theropods to birds is modelled as a correlated progression of functionally linked traits, exactly as in the case of the origin of mammals, with much the same traits at issue.

The environmental circumstances, ecological significance, and what drove the lineage leading to birds were discussed in Chapter 6, section 6.4.

8.3 Tetrapods

The modern lungfishes (Dipnoi), coelacanth (Actinistia), and tetrapods (Tetrapoda) constitute an unchallenged monophyletic taxon Sarcopterygii, although the relationships amongst these three sub-taxa are not confidently resolved either by molecular or morphological evidence. Different analyses have supported both a dipnoan–coelacanth and a dipnoan–tetrapod relationship (Zardoya et al. 1998; Takezaki et al. 2004). The most recent investigation to date utilized a huge dataset of 1290 nuclear genes and supported the dipnoan–tetrapod relationship (Liang et al. 2013). Whether correct or not, this study does not detract from the conclusion that the three lineages diverged from what amounts to a virtually irresolvable trichotomy, the internal branch lengths being a great deal shorter than the external branch lengths leading to the three modern clades.

Wherever its root may lie, the tetrapodomorph stem lineage that diverged from the common ancestor of sarcopterygians and culminated in the tetrapods includes fossils of several intermediate grades between 'fish'-grade tetrapodomorphs and crown tetrapods (Fig. 8.19). For many years, the most basal limb-bearing tetrapodomorph known in any detail was *Ichthyostega* (Jarvik 1996). However, since the description of *Acanthostega* in the early 1990s, the morphological gap has continued to be filled by the discovery of around a dozen possible stem taxa (Clack 2012). Only a few of these taxa are relatively complete and well known, and most are still represented by fragmentary material, often previously misidentified in collections as sarcopterygian fish. Research in this field is rapid, however, and many more details bearing on the origin and early history of tetrapods can be expected. The important information that is accruing throws light on both the sequence of acquisition of tetrapod characters, and on the environmental conditions in which it occurred.

8.3.1 The grades of fossil stem tetrapods

Eusthenopteron

The lobe-finned fish *Eusthenopteron* (Fig. 8.17a) occurs in Late Devonian deposits, and has been taken as a model for the 'fish' ancestor of the tetrapods ever since it was first described by Whiteaves (1883). It possesses numerous characters relating it to tetrapods (Andrews and Westoll 1970; Jarvik 1980). Of particularly striking note is the form of the internal skeleton of the paired fins. These consist of the easily recognized homologues of the tetrapod humerus, radius, and ulna in the pectoral fin and femur, tibia and fibula in the pelvic, although they still support fin rays in a large fin membrane and there are no digits. The presence of internal choanae, and many aspects of the skull structure and bone pattern are also clearly homologous with tetrapod characters.

Notwithstanding these features shared with tetrapods, *Eusthenopteron* was undoubtedly a fully aquatic animal, adapted for ambush predation to judge from its pike-like proportions and posterior position of the median fins. While presumably a facultative air breather, it nevertheless possessed a full complement of well-developed gills. Indeed, it was no more than a derived member of a considerable diversity of completely 'fish'-grade tetrapodomorphs dating from the most basal form which is the Early Devonian Chinese *Kenichthys* as reviewed by Clack (2012).

Panderichthys

The late Middle Devonian of Latvia has yielded specimens of the tetrapodomorph *Panderichthys* (Fig. 8.17b), which combines a generally *Eusthenopteron-like* structure with a number of incipient tetrapod characters of skull, body, and fins. Both the dorsal and the ventral regions of the body are covered by rhomboid-shaped, slightly overlapping scales arranged in oblique rows, much as in other basal tetrapodomorphs (Witzmann 2010). Compared to *Eusthenopteron*, the body is flattened, giving it a somewhat crocodile-like appearance, and it lacks median fins (Vorobyeva and Schultze 1991). The skull has evolved the flattened form characteristic of early tetrapods, and the orbits face dorsally. The preorbital region is elongated with paired frontal bones and the postorbital length relatively reduced, which is associated with a reduction of the cranial kinesis between the skull table and the rest of the skull found in more basal taxa. Surprisingly perhaps, the braincase does retain the ancestral intracranial joint between an anterior ethmosphenoid part and a posterior

oticooccipital part of the braincase that is found in *Eusthenopteron* (Ahlberg et al. 1996). Brazeau and Ahlberg (2006) described the anatomy of the spiracular region as intermediate between that of *Eusthenopteron* and that of tetrapods. The palatoquadrate or upper jaw is much shallower in accordance with the flattening of the skull. The hymandibula bone still attaches to the opercular cover of the gill cavity distally, although it does not extend further ventrally and is therefore shorter, straighter, and more like the stapes of tetrapods. The spiracle runs between the palatoquadrate and hyomandibula and is wider and straighter, which suggests that it had an enhanced role in inspiration during aquatic respiration, as occurs in modern bottom-adapted elasmobranchs; it is very unlikely that the hyomandibular and spiracular pouch had evolved the function of airborne sound reception at this stage.

The paired fins retain fin rays, but the skeleton of the pectoral fin is more tetrapod like in some respects. The scapulo-coracoid of the shoulder girdle is larger and the fin no longer extends strictly backwards, but has a ventro-lateral inclination, while the glenoid faces postero-laterally. The humerus is more slender than that of *Eusthenopteron*, compressed dorso-ventrally, and it shows evidence of greater differentiation of the musculature into dorsal, extensors, and ventral flexors (Boisvert 2009). There are two separate facets for articulation between the humerus and the ulna and radius, rather than a single one. While there are no digits as such, there is a smaller number of elements distal to the radius and ulna, four of which are arranged as an arc around the end of the ulna. These were interpreted by Boisvert and colleagues (2008) as radial bones that have essentially the same relationship to the ulna as do the digits found in *Acanthostega*. Therefore, although fin rays are still present rather than discrete digits, the fore limb of *Panderichthys* does represent a stage in the evolution of the definitive tetrapod limb. The hind limb is less modified (Boisvert 2005). The pelvis is very small and there is no development of an iliac process. There are only five bones in the endoskeleton, identifiable as the

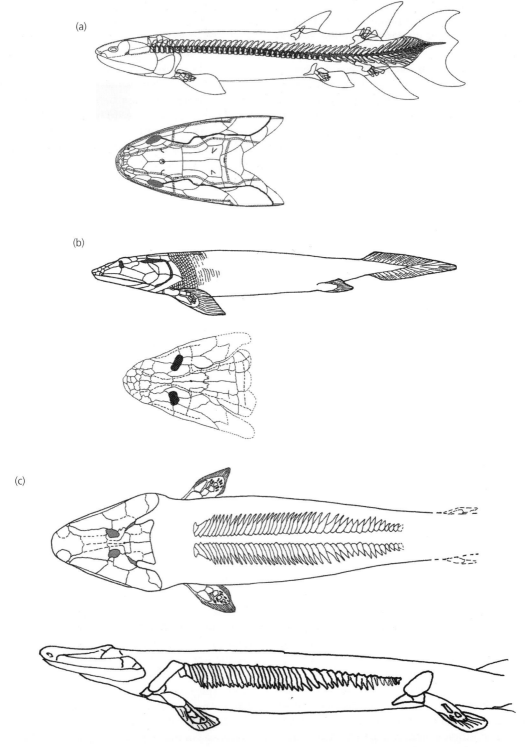

Figure 8.17   Finned tetrapodiforms. (a) *Eusthenopteron.* (b) *Panderichthys.* (c) *Tiktaalik.* ((a) from Andrews and Westoll 1970 reproduced with permission from Cambridge University Press; (b) from Clack 2012; (c) from Clack 2012, as modified from Daeschler et al. 2006.)

femur, tibia, fibula, intermedium, and fibulare, and they appear to have been tightly connected to one another, allowing very little flexibility. Functionally, the fin structure of *Panderichthys* indicates that the pectoral fins could bear the weight of the animal, presumably being capable of holding the head clear of the water for aerial breathing, but the pelvic fins could not. The latter could, however, reach the ventral surface and so were large enough to act as pivots and play a role in locomotion by lateral undulation of the body axis in contact with the substrate (Boisvert 2005; Clack 2009).

Tiktaalik

The discovery in Arctic Canada of articulated specimens of the tetrapodomorph *Tiktaalik* (Fig. 8.17c) was a landmark event in the study of the origin of tetrapods (Daeschler et al. 2006), for in a number of respects its morphology lies between that of *Panderichthys* and a limbed tetrapods such as *Acanthostega*. Like *Panderichthys*, the body was dorso-ventrally flattened and crocodile-like in form, the skull also flat with dorsally facing eyes, and the intracranial joint still unfused. However, the snout was longer than the postorbital part of the skull length although less so than in *Acanthostega*. The braincase still had an unfused intracranial joint. At the rear of the skull the spiracular notch is wider than that of *Panderichthys* and there are no longer opercular bones, which implies greater reliance on air-breathing. On the other hand, the gill arches are still much better developed than in *Acanthostega* indicating that gill respiration was still the principle source of oxygen. The hyomandibula is intermediate in structure and orientation, being shorter, straighter, and horizontally arranged, although not so extensively modified as in *Acanthostega*. The palate is flat and horizontally oriented and the basipterygoid articulation between the palate and braincase enlarged, as in the tetrapods. These modifications to the skull structure, including the change in the proportions of the snout and skull table, point to a reduction and perhaps a complete loss of the ability to feed by suction as, typically, in fish and the adoption of the biting mechanism of tetrapods (Downs et al. 2008).

The postcranial skeleton is also modified in a number of significant ways that presage the tetrapod locomotory system. The vertebral column is associated with relatively well-developed, overlapping ribs implying a role in weight support while the animal was partially or completely out of water. The extrascapular bones behind the head have been lost so there is no longer the bony connection between the head and the shoulder girdle. This creates for the first time a degree of neck flexibility. The scapulo-coracoid of the shoulder girdle has expanded dorsally and ventrally, and the glenoid fossa for articulation of the humerus faces even more laterally than in *Panderichthys*, indicating a wider range of movements of the front appendage. The appendage itself is still technically a fish fin because there are fin rays rather than digits. However, the distal endoskeletal elements, including wrist bones, are more robust and the fin rays are stouter and reduced in length, indicating that the pectoral fins were better able to support the animal in shallow water and perhaps also on land (Shubin et al. 2006). The pelvic girdle and fin is also more progressive than in other finned tetrapodomorphs (Shubin and Daeschler 2014). The pelvis is very much larger and extends dorsally and anteriorly. However, it still lacks a direct attachment to the vertebral column. The fin is only partially known, but judging from the parts preserved was clearly also very much enlarged, indicating that it too was involved in substrate locomotion.

Ventastega

Fragmentary remains of *Ventastega* (Fig. 8.18a) occur in the Upper Devonian of Latvia (Ahlberg et al. 2008). Most of the skull is known, although from separate fragments, while of the postcranial skeleton the shoulder girdle and part of the pelvis are the only parts so far described. Nevertheless, as far as it is possible to tell, *Ventastega* is almost ideally intermediate in structure between *Tiktaalik* on the one hand and the more derived form *Acanthostega* on the other, such as in the proportions of the skull, in which the skull table is relatively shorter and the eyes larger than those of *Tiktaalik*. The braincase is more like that of *Acanthostega*, notably in possessing a fenestra vestibuli in the side wall, which implies that the upper part of the hyomandibular was already transformed into a sound-conducting stapes. On the other hand, the interorbital wall of *Ventastega* is solid, like that of *Tiktaalik* (Downs et al.

2008) and *Panderichthys*. The pectoral girdle has a generally tetrapod form that is quite different from that of *Tiktaalik*, leading Ahlberg and colleagues (2008) to infer that a tetrapod limb with digits had evolved. One interesting difference from both *Tiktaalik* and *Acanthostega* is the presence of a very large spiracular notch, perhaps associated with a greater dependence on air-breathing than the former. Alternatively, of course, it may simply represent an apomorphic adaptation for a particular habitat.

Acanthostega

Although discovered long ago in Upper Devonian rocks of East Greenland, along with *Ichthyostega* (Säve-Söderbergh 1932), *Acanthostega* (Fig. 8.18c) remained poorly known until Jennifer Clack came across a number of specimens obscurely stored in the collections of the British Antarctic Survey in Cambridge, and subsequently collected a number of beautifully preserved specimens from the original locality (Clack 1988). Since then, the anatomy of *Acanthostega* has been described in great detail, demonstrating its position as an almost ideal basal tetrapod possessing a combination of ancestral and derived tetrapod characters (Clack 1994, 1998, 2012; Coates 1996).

The whole animal is around 1 m in body length, and the skull flat with dorsally facing orbits as in *Ventastega*. The nature of the dorsal dermal scales is not known, but the ventral region is covered by obliquely oriented gastral scales arranged in a chevron pattern, and which therefore resembles the typical basal tetrapod pattern rather than the primitive rhomboidal pattern of *Panderichthys* (Witzmann 2010).

The skull roof still has a number of bones that were lost in later tetrapods, notably a preopercular bone at the back. The braincase is very tetrapod-like, having lost all signs of the intracranial joint, although there is no occipital condyle at the back of the skull for attachment to the vertebral column, and the notochord of the column still continues forwards within the braincase base. The hyomandibular has achieved the tetrapod condition, being even shorter than in *Ventastega* and with a footplate clearly attached to the fenestra vestibuli of the braincase wall (Clack 2006); presumably by this stage airborne sound reception had become an important sense. At the back of the head there is still

a full complement of well-developed branchial arches, and the grooves running along their length indicate a good blood supply and therefore a continued high level of reliance on gill breathing. The spiracular chamber is large and widely open externally, and there is some argument about the significance of its structure (Clack 2006). It may have been a means of increasing the intake of water to the gills appropriate in a bottom-living animal, as was suggested for *Panderichthys* and which would correlate with the well-developed gills. Alternative suggestions are that it was an adaptation for air-breathing, or that even at this early stage it was sealed off as an air-filled cavity associated with sound reception, a role it was destined to adopt at some point in tetrapod evolution.

The vertebral column of *Acanthostega* indicates an aquatic animal supported by water rather than one adapted to terrestrial locomotion (Coates 1996). The notochord is prominent and the vertebral elements surround it fairly superficially. The zygapophyses between adjacent vertebrae are small, as are the ribs all along the length of the vertebral column. At the hind end, the column extends into a large tail fin supported by fin rays and presumably the most important structure for actual swimming. The most remarkable anatomical feature of *Acanthostega* proved to be the limbs (Coates 1996), especially the replacement of fin rays by terminal digits. There are eight of these, front and back, forming an arc distal to the epipodials. However, the limbs are still short and paddle-like. The humerus is very broad and the radius is longer than the ulna; in fact, the arrangement of these three bones in *Acanthostega* is less tetrapod-like than in *Tiktaalik*. There is no ossified wrist, and the limbs could hardly have borne the weight of the animal out of water. This apparent paradox of tetrapod limb features but overall a weak fore limb structure led to the suggestion that, even at this early stage in tetrapod evolution *Acanthostega* might be secondarily aquatic (Ahlberg 2011). This may also be indicated by the shoulder girdle which, though enlarged compared to more basal tetrapodomorphs, is still small for a tetrapod, and the lateral-facing glenoid is flat rather than concave. The hind limbs are supported by a pelvis that is also intermediate in size between those of more basal tetrapodo-morphs and later tetrapods. They differ from the fore limb in having the epipodials, tibia and fibula, of equal length and in possessing a number of ossified tarsal bones in the ankle. Taken together, the limbs appear to be associated with shallow aquatic locomotion, acting as pivots for lateral undulatory locomotion on a soft substrate, as paddles during swimming, or in the case of the fore limbs perhaps as props to hold the head out of the water during air breathing. Presumably all three modes of action were possible (Pierce et al. 2013).

Despite the tetrapod features of the skull and postcranial skeleton, *Acanthostega* was clearly a mainly if not entirely aquatic animal, with well-developed gill breathing as well as accessory air breathing. The limbs were designed for a form of aquatic locomotion and it is doubtful if the animal could have seriously ventured onto land at all. Whether the changes in the morphology of the skull relate to a shift from suction feeding as in typical fish to grasping feeding as in tetrapods is debatable. The loss of the intracranial mobility and form of the sutures between the cranial bones suggest a biting mode (Markey and Marshall 2007), but Anderson and colleagues (2013) argued from the conservative structure of the jaw that *Acanthostega* still had an essentially primitive, fish-like feeding mechanism.

Ichthyostega

From the initial announcement (Säve-Söderbergh 1932) to the very detailed description by Jarvik (1996), the Upper Devonian East Greenland *Ichthyostega* (Fig. 8.18b) was regarded as an archetypical intermediate between fish and tetrapod. As well as limbs with digits, it had tetrapod skull proportions, a vestigial operculum, and at the time was mistakenly believed to have lacked branchial arches. On the other hand, it had retained a fish-like tail complete with fin rays, and lateral lines in canals within the skull bones, rather than in superficial grooves and, as is now known, grooved branchial arches indicating the presence of gills, albeit reduced. However, new material plus appreciation that *Acanthostega* is of a similar grade but less specialized has somewhat demoted the

significance of *Ichthyostega*. The braincase and ear region are unlike those of any other known vertebrate (Clack et al. 2003), notably in having a much larger otic region and with a large cavity alongside and above it, which was possibly an air-filled structure capable of detecting water-borne sound. The vertebral column is also unique. Unlike the condition found in *Acanthostega*, it is differentiated into neck, trunk, and lumbar regions (Ahlberg et al. 2005). The individual vertebrae are weakly developed but the ribs are extremely large and overlap one another by means of costal plates, in stark contrast to any other early tetrapods. Presumably this massive rib cage served as a support for the body, perhaps while out of water, rather than for the costal breathing associated with much later terrestrial tetrapods (Clack 2012). The shoulder girdle is large and the fore limb stoutly built, indicating a weight-bearing capacity. The pelvis is also larger than in *Acanthostega* and attaches via a sacral rib to the vertebral column. However, the hind limb is smaller than the fore limb, perhaps as little as half the length, and according to the reconstruction by Ahlberg and colleagues (2005), it was oriented posteriorly and therefore was unlikely to have been able to support the body weight out of water.

While it is clear that *Ichthyostega* had acquired additional tetrapod characters not present in *Acanthostega*, including further reduction of the gills in the adult and a reduction in the number of digits from nine to seven, it was a highly specialized animal that needs to be treated with caution as an illustration of a grade in the evolution of tetrapods. In particular, the locomotion probably resembled the 'crutching' locomotion of mud skippers or phocid seals, in which the fore limbs are used synchronously to raise the front of the animal and push it forwards. The hind limbs were unable to reach the substrate at all and appear to have been functionally paddles used for aquatic locomotion, in conjunction with the large tail fin (Pierce et al. 2012; Pierce et al. 2013).

Tulerpeton

Tulerpeton (Fig. 8.18d) is from the Upper Devonian of Russia. The most important specimen is an

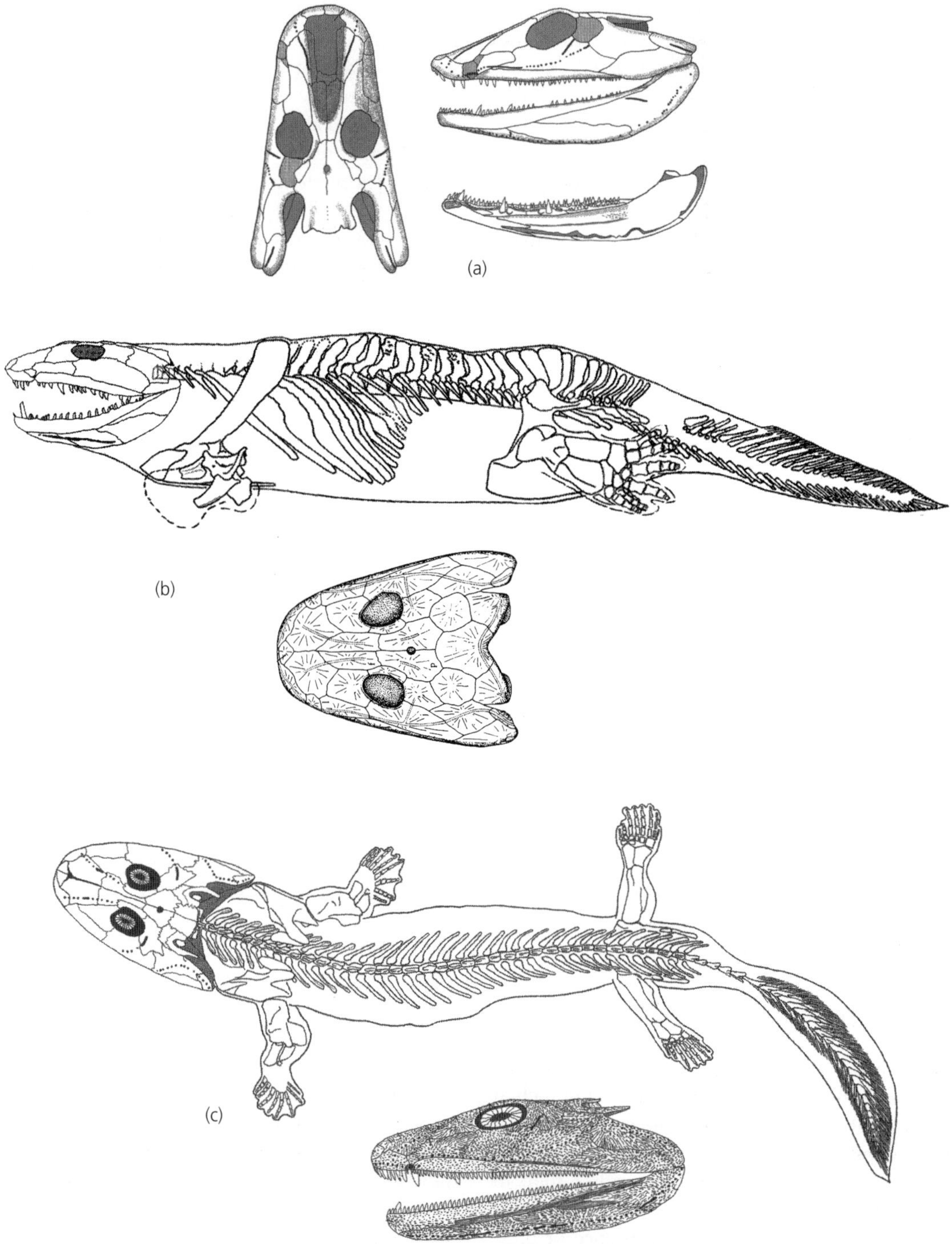

Figure 8.18 Limbed tetrapodiforms. (a) *Ventastega.* (b) *Ichthyostega.* (c) *Acanthostega.* (d) *Tulerpeton.* ((a) from Ahlberg et al. 2008; (b) above from Clack 2012, below from Carroll 1988; (c) and (d) from Clack 2012. Courtesy of Indiana University Press. All rights reserved.)

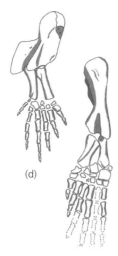

(d)

Figure 8.18 (*continued*)

incomplete skeleton including a fore and a hind limb, shoulder girdle, and dermal scales characteristic of tetrapods (Lebedev and Coates 1995). Other associated specimens are a partial pelvis and a number of isolated skull and vertebral fragments (Lebedev and Clack 1993). Thus, it is not yet very well known, but nevertheless is an important taxon because the limb structure is more derived than that of the other Devonian tetrapods. The limbs are relatively better developed and in particular the epipodials, radius and ulna in the fore limb and tibia and fibula in the hind limb, are longer and more gracile. The number of digits is down to six. The cranial bones reveal little about the structure of the skull, and it is not kown whether branchial arches were present.

Based on this limited material, *Tulerpeton* is more crownward than *Acanthostega* and *Ichthyostega*, and more specimens may prove very important. One unexpected aspect is that the locality where it has been found is marine, and indeed probably a long way offshore, which has implications for the ecological circumstances and adaptations of early tetrapods.

Crown tetrapods

Recent cladistic analyses agree that all the post-Devonian tetrapods constitute a monophyletic group Tetrapoda, and since by most accounts it includes the ancestry of both modern amphibians and amniotes, at least the great majority are members of the crown group too. From the very beginning of the Carboniferous, tetrapods commenced an explosive radiation into many taxa whose interrelations certainly are a matter of continuing analysis and dispute. It is often difficult to infer what was the ancestral tetrapod condition because of early diversification of, for example, vertebral structure and patterns of dermal bones of the skull. Characters which are probably diagnostic of the crown group Tetrapoda include more robust vertebrae with substantial zygapophyses between them, and bearing elongated ribs. The first vertebra, the atlas, articulated with an occipital condyle at the back of the skull, and along with the second, the axis, permitted more extensive head movements. The scapula-coracoid enlarged, and the pelvic girdle evolved three ossifications, ilium, ischium, and pubis. The limbs are enlarged and fully weight-bearing, and the digits reduced to five. The tail has lost the fin rays. In the skull, the external nostril enlarged, and there was a tendency to reduce the palatal fangs and enlarge the marginal teeth of the jaws, although this is a continuation of a trend seen in the more basal tetrapods and no precise condition can be defined for the crown group.

8.3.2 The pattern of acquisition of tetrapod characters

Between the arbitrary start and end points represented by the *Eusthenopteron* grade and crown tetrapod, major evolutionary changes in characters associated with virtually all the functions of the animals occurred. These include changes in the skull morphology related to radical modification of the feeding mechanics. The ancestral fish condition, which consists primarily of suction by kinetic expansion of the oral and branchial cavities, transformed into the tetrapod mode where the skull is akinetic and the jaws generate large biting and grasping forces. The process was accompanied by changes in the dentition and extensive modifications to the jaw musculature. The loss of the orobranchial kinetic system, along with the branchial arches and opercular apparatus, also reflects the shift in the respiratory method from air breathing as an accessory mode to the obligatory mode it has in adult tetrapods involving a ribcage capable of costal ventilation. Modifications to the skull also indicate that the senses changed from an exclusively aquatic context to reception of aerially transmitted information. Olfaction has potentially far greater sensitivity and range in air, and therefore benefits from the increase in the olfactory epithelium in the enlarged snout and external nostril. Vision requires a new accommodation mechanism. Hearing in air requires the sound-conducting property of the reduced hyomandibular and open fenestra ovalis, with or without an air-filled tympanic cavity.

The postcranial skeleton was radically changed from a design for typical pisciform swimming by lateral undulation of the body using a caudal fin to terrestrial locomotion. The vertebral column evolved a greater load-bearing ability by changes in the vertebral structure and elaboration of the ribs, whilst the head became more flexibly attached. The scapulo-coracoid and pelvic girdle both enlarged, the former becoming well attached to the ribcage by muscle and the latter evolving a direct bony articulation with the sacral region of the vertebral column for thrust transfer. The propulsive limbs became large and muscular enough to move a relatively massive organism on land, and digits replaced fin rays for traction on the ground.

These evolutionary modifications in feeding, olfaction, hearing, and locomotion are reflected to at least some extent in the skeletal morphology. Accompanying physiological transitions do not generally affect skeletal morphology and so are invisible in the fossils, but by implication from the change in habitat and in comparison with modern relatives, they must have included tolerance of the greater extent of temperature fluctuation, and mechanisms to control the osmotic and ionic imbalances to which terrestrial organisms tend to be subject. Superimposed on all the changes, the circulatory system and central nervous systems must have been greatly modified to serve and control this complex of new structures and functions.

The tetrapodomorph fossils between *Eusthenopteron* and the crown tetrapod grades throw a certain amount of light on the way in which this complex of significant changes occurred in the evolving lineage (Clack 2012).

The transition from *Eusthenopteron* to *Panderichthys* grades involved a feeding mechanism change shown by the loss of cranial kinesis; a shift in ventilation towards reduced reliance on the aquatic mode shown by the reduction of the opercular region; modification of the locomotion towards shallow-water, substrate-contact propulsion shown by the flat body, loss of medial fins, and enlarged humerus; possible evolution towards aerial sensory modalities shown by the relatively longer snout, large, dorsally oriented eyes, and reduced hyomandibular.

In the transition to the *Tiktaalik* grade, relative increase in preorbital length implies a modest change in the feeding mechanism, while loss of the opercular bones suggests increased reliance on air breathing despite the retention of well developed branchial arches. The connection between the back of the skull and the shoulder girdle is lost, and the enlarged scapulo-coracoid and fore-limb bones, including the wrist, indicate greater substrate locomotory ability. The hyomandibular lost the process that had connected it to the operculum, perhaps increasing its sensitivity to airborne sound.

The evolution from *Tiktaalik* to *Acanthostega* grades continued these trends. The snout was relatively even more elongated so the skull had acquired more or less typical tetrapod proportions and the two parts of the braincase were finally consolidated into a single unit. These changes, along with the modified pattern of cranial sutures imply a tetrapod biting mode of feeding (Markey and Marshall 2007). Ventilation still involved well-developed branchial arches, though less so than in *Tiktaalik*, indicating a greater relative role for air breathing. The locomotory mechanism was the most modified function at this step. Along with the loss of the attachment of the head to the shoulder girdle, the enlarged and modified atlas and axis vertebrae ensured the beginning of a flexible neck. The vertebral column was relatively little modified, apart from the ribs now attaching to the neural arches. Further enlargement of the scapulo-coracoid, pelvis including attachment by a sacral rib, limb bones, and most significantly, replacement of the fin rays by eight digits point to more effective substrate locomotion in shallow water. Possibly *Acanthostega* was capable of some degree and form of terrestrial locomotory ability, though unable to completely support its weight on the limbs (Pierce et al. 2013). In the ear region, the head of the further reduced hyomandibular lies within a fenestra vestibuli, a clear sign of aerial hearing.

Ichthyostega, having been shown to be highly aberrant in several features, should be taken only with caution as representative of a grade between *Acanthostega* and crown tetrapods. Indeed, its phylogenetic position at this point is not completely secure. Thus it adds very little if anything to a clear picture of the transition to tetrapods. Leaving it aside, the final shift from the *Acanthostega* grade to crown tetrapods is possibly bridged by the currently poorly known *Tulerpeton*, and included further consolidation of the skull, palate, and occiput into a more robust structure, and loss of the branchial arches in the adult, although their retention as a paedomorphic condition was common. Enlargement of the external nostril indicates obligate air breathing. The dentition changed by reduction and loss of palatal tusks leaving the marginal teeth as the most prominent. The postcranial skeleton illustrates the appearance of fully expressed tetrapod locomotion, associated with robust vertebrae articulated by well-developed zygapophyses, and enlarged girdles and limbs.

Other early stem tetrapod taxa mentioned, notably *Ventastega* and *Tulerpeton*, are not as yet known in sufficient detail to add much to this outline view of the sequence of acquisition of crown tetrapod characters, but in due course will no doubt contribute significantly. There are several other tetrapodomorphs represented by material that at present is even more fragmentary, such as the jaw of the Australian *Metaxygnathus* (Campbell and Bell 1977), jaw and postcranial scraps of *Elginerpeton* from Scotland (Ahlberg 1995), and a humerus intermediate in structure between *Tiktaalik* and *Acanthostega* that dates from the Famennian stage and was found in Pennsylvania (Clack 2005, 2009).

There is plenty more detail of the story to come, and the taxa known so far also serve as a reminder that the fossils are not a sequence of actual ancestors, but specialized relatives of ancestral grades.

The path from fish to tetrapod certainly was not a monotonic trend, but part of a branching tree from which a sequence of hypothetical ancestral stages has been somewhat arbitrarily lined up. Within these constraints, the general picture to emerge is that at each stage represented by fossil material, evolutionary changes in characters associated with the different functional systems are seen to have occurred, and never change in any one of them exclusively. This pattern is unambiguously in accord with the assumption of the correlated progression model that the different parts and functions of the organism such as feeding, ventilation, locomotion, and sense organs are relatively tightly integrated, and so no one of them can evolve to a significant extent without accompanying evolutionary modifications to others (Fig. 8.19).

Evidence bearing on the environmental circumstances associated with the correlated changes that resulted in the origin of tetrapods is adduced and discussed in Chapter 6, section 6.3.

8.4 Turtles

The chelonians famously have amongst the most extreme modifications of the tetrapod body plan. The presence of the shell, carapace above and plastron below, is correlated with radical modifications to the axial and appendicular skeleton associated with locomotion and the ventilation mechanism. The skull has lost the ancestral temporal fenestration, several cranial bones, and all trace of the dentition, while the ear region has been remodelled by independent evolution of a tympanic membrane and highly compliant stapes. Indeed, the necessary reorganization of the entire body plan in what must have been a coordinated way has been considered a candidate for some kind of macromutational event, a 'hopeful monster', such as the effect of a single regulatory gene (Rieppel 2001). However, a small number of fossil grades of stem turtles have been described and, while far from fully elucidating the sequence of evolutionary changes from a basal amniote, do show that the transition was not quite such an incomprehensible event.

Traditionally chelonians were thought of as more or less living fossils, tracing their ancestry back to

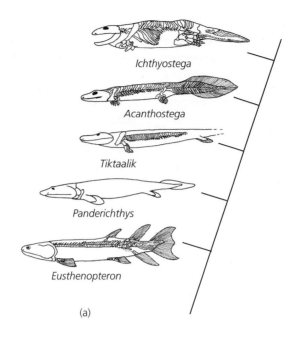

(a)

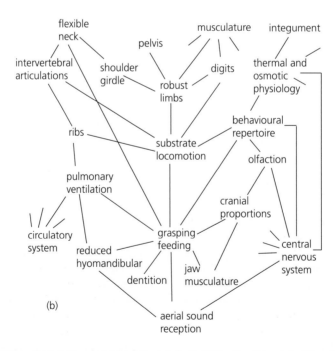

(b)

Figure 8.19 (a) Sketch cladogram of principal known intermediate-grade fossils. (b) Functional interrelationships of main morphological parts and physiological functions of the ancestral tetrapod. (From Kemp 2007.)

early amniotes with unfenestrated skulls, and therefore diverging before the synapsids or diapsids had evolved. Most morphological-based phylogenetic analyses associated them with the Parareptilia, a group of amniotes that is the sister-group of the rest of the reptiles. There were differences about exactly which parareptilian taxon is closest to Chelonia, with Lee (1997) championing the Permian group of large forms Parieaisauria, and Laurin and Reisz (1995) the smaller, somewhat lizard-like Procolophonia of the Triassic. Around the same time, the egregious conclusion was reached by deBragga and Rieppel (1997) that chelonians are the sister-group of the marine reptile taxon Sauropterygia, and that the two together constitute a clade within Diapsida, to the exclusion of parareptiles altogether. Perhaps surprisingly, subsequent molecular evidence has also generally supported a relationship of turtles with diapsids, although whether with the archosauromorphs in particular (Shedlock and Edwards 2009), with the lepidosauromorphs, or as the sister-group of the diapsids as a whole is not yet resolved (Lyson et al. 2011). In fact, the exact relationships of the group to other living amniotes is virtually immaterial as far as deciphering the morphological transition from ancestral amniote to chelonian is concerned, since chelonians are so extremely derived in so many respects from all other amniotes.

There are presently three relevant fossils which throw some light on the acquisition of chelonian characters (Lyson et al. 2011). Which is the most basal grade depends on which phylogenetic hypothesis is followed. That of Lee (1996, 1997) has pareiasaurs at the base, that of deBragga and Rieppel (1997, Rieppel and Reisz 1999) something like a placodont, and a weak case can be made for either. However, the most convincing candidate for the basalmost stem chelonian is *Eunotosuraus*, while *Odontochelys* is the first undisputed stem-chelonian.

8.4.1 The grades of fossil stem chelonians

Eunotosaurus

Dating from the Middle Permian of South Africa, the peculiar animal *Eunotosaurus* (Fig. 8.20b) has figured in discussions of the origin of turtles ever since the original description by Watson (1914)

and the detailed comparisons of Cox (1969). The main chelonian characters are the reduction of the presacral vertebrae to ten, coupled with broadly expanded ribs that are T-shaped in section and certainly resemble those of chelonians. Although there is no trace of any separate dermal armour associated with the ribs, the texture of their dorsal surface is roughened in a fashion reminiscent of dermal tissue (Lyson et al. 2011). Despite the extreme shortening of the trunk and widening of the body, the limbs of *Eunotosaurus* are much like those of other parareptiles. The skull is generally short and broad in shape, with a full marginal dentition and a complete set of amniote dermal bones including those that were subsequently lost in turtles.

Odontochelys

Odontochelys (Fig. 8.20a and c) is a Late Triassic chelonian from China with an incomplete shell (Li et al. 2008). That it is a stem chelonian is not in doubt, since it possesses a fully formed plastron constructed from the same pattern of dermal plates as turtles. However, the carapace is very incompletely formed, consisting only of a series of neural plates which were attached to but were not fused with the neural spines of the ten dorsal vertebrae. The ribs are broadened, but not expanded into costal plates as in crown turtles. The limbs and girdles are essentially chelonian but retain a number of ancestral features and there is still a relatively long tail. The skull has also retained some ancestral characters. The postorbital region is not shortened, and there are small, sharp teeth along the premaxilla, maxilla, and dentary, in addition to palatal teeth.

Proganochelys

This European Upper Triassic form (Fig. 8.20d) has evolved the fully developed shell (Gaffney 1990). The carapace has a complete set of dermal neurals fused to the dorsal vertebrae, pleurals fused to the ribs, and marginals. The plastron, as already evolved in *Odontochelys*, is similarly complete. However, the limb girdles at this stage are now enclosed entirely within the shell, and the characteristically chelonian mode of limb-based locomotion had evolved. The tail is reduced. The skull also has new chelonian characters, including loss of the marginal teeth, although small teeth on the palatal

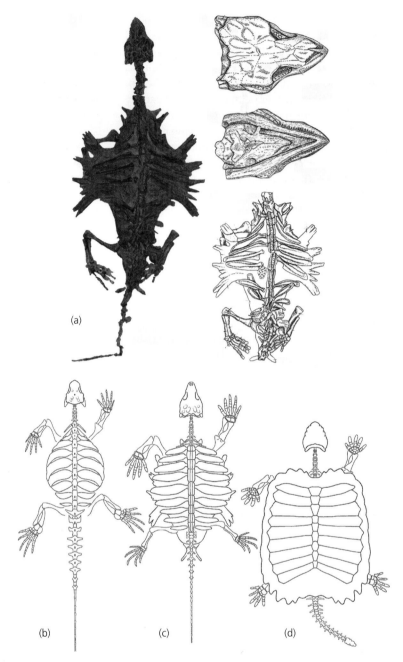

Figure 8.20 Stem turtles. (a) Fossil in dorsal view and reconstruction of the skull of *Odontochelys*. (b) *Eunotosaurus*. (c) *Odontochelys*. (d) *Proganochelys*. ((a) from Li et al. 2008; (b)–(d) from Lyson et al. 2011 reproduced with permission from the Royal Society.)

bones have been retained. The posterior part of the skull is remodelled along chelonian lines by loss of the postfrontal and postparietal bones, and appearance of a prominent otic notch of the quadrate and surrounding bones, presumably to support a chelonian tympanic membrane.

8.4.2 The pattern of acquisition of chelonian characters

Lee (1996, 1997) discussed the pattern of acquisition of chelonian characters, but his study was constrained by his phylogenetic analysis that had turtles nested within the pareiasurs, a view that has not found universal favour. This forced him to the conclusion that the first stage in shell evolution was a row of small osteoderms along the dorsal midline that were not attached to the neural spines. This was followed by the evolution of a complete covering of the dorsal surface by osteoderms, and finally these united to constitute a continuous dorsal carapace. By this stage too, lateral undulation of the vertebral column was restricted and modifications of the limbs and girdles were necessary as the locomotory method shifted to an entirely limb-driven mode with no longer any lateral undulatory component. However, he did not consider *Eunotosaurus*, and had not the benefit of *Odontochelys* at that time. In contrast, Lyson and colleagues (2011) did consider these two forms, which led them to conclude that the initiation of the evolution of the shell consisted of a reduction of the dorsal vertebral number to ten in association with expansion of the ribs. This was followed by completion of the plastron as in *Odontochelys*. The evolution of the carapace, incorporating osteoderms attached to the vertebrae and ribs, was the final stage, as it is not known until the grade represented by *Proganochelys*.

The fossil evidence is not yet comprehensive enough to judge between these two hypotheses for the origin of the turtle shell, nor to add much detail about the concurrent evolution of the skull. Both pareiasurs and placodonts, as well as *Eunotosaurus* possessed shortened, robust skulls indicating a herbivorous habit, and thus all could, with a little imagination, be regarded as models for the ancestry of

the chelonian cranium. However, even this limited information shows that the evolution of the highly modified postcranial skeleton required correlated shifts in the vertebral column, ribs, and osteoderms, simultaneously with the shortening of the limbs and shift in the anatomical position of the shoulder girdle relative to the axial skeleton. Modification to the ventilation mechanism must necessarily have occurred hand in hand with the increasingly immobile ribcage, and the process is also accompanied by the adoption of short, more powerful jaws adapted, it seems, for a herbivorous diet. Taken together, the picture emerges of a lineage evolving a radically new mode of life in which fast locomotion is traded in favour of physical protection from predation. Correlated with the loss of speed was the adoption of a diet that no longer required active hunting. Correlated with the protective shell was the modification of the ventilation system by replacing costal breathing by movements of the limb girdles within the carapace. In connection with the shortened, strengthened, akinetic skull and remodelled jaw muscles for herbivory, there was the potential for an enlarged, well-supported tympanic membrane. All these different changes must be viewed as part of a correlated progression (Fig. 8.21), and the characters of the few stem-chelonians so far known offer support for such an interpretation.

As discussed in Chapter 5 (section 5.3), a good deal of effort is being expended on the development of chelonians in an attempt to further understand the evolutionary transition to such a bizarre version of the amniote organism.

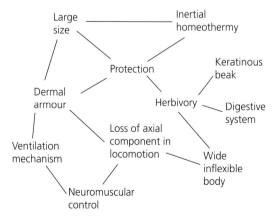

Figure 8.21 Functional interrelationship of some of the major integrated aspects of a turtle.

8.5 Cetacea

Of all the mammalian orders, the two which qualify most indisputably as higher taxa in their own right are the bats and the whales. The fossil record of the origin of bats is extremely poor, consisting of no more than certain Eocene species with relatively minor plesiomorphic characters retained. However, that of whales is much more revealing, thanks to a series of limbed Eocene stem cetaceans from Pakistan and India. This case is also of special interest as the only significant sequence of grades from fully terrestrial to obligatorily marine tetrapods, an evolutionary route also followed at various times in the past by many taxa of reptiles, the sirenian mammals, and to a large though not complete extent, penguins and great auks amongst the birds.

The phylogenetic position of the Cetacea within placental mammals is no longer in any doubt as a result of numerous molecular studies and a huge database of analysed sequences. They comprise a monophyletic group nested within the Artiodactyla, the even-toed ungulates, as the sister-group of the hippopotamuses. Indeed, the whole order is properly referred to as the Cetartiodactyla (Spalding et al. 2009). Arguments based on morphology that whales were not related to artiodactyls but instead to the extinct carnivorous taxon Mesonychia were dispelled with the discovery in the late 1990s that the Eocene stem cetaceans still possessed functional limbs with the highly characteristic artiodactyl structure.

8.5.1 The grades of fossil stem cetaceans

Indohyus

There are several grades of limbed cetaceans known, with varying degrees of limb reduction. The most basal may the raoellids such as *Indohyus* (Fig. 8.22a), which was a raccoon-sized artiodactyl with unreduced limbs, but that had dense bones and also an oxygen isotope ratio of the bones suggestive of animals spending much of its time in water (Thewissen et al. 1996). The molar teeth are essentially primitive artiodactylous in form, but have a pattern of wear similar to that in other, more derived stem cetaceans. The tympanic bulla also has a whale-like thickened lip suggesting adaptation for hearing under water.

Pakicetids

The wolf-sized Middle Eocene form *Pakicetus* (Fig. 8.22b) was first recognized as a basal cetacean by similarities of its dentition to that of better known Upper Eocene primitive whales (Thewissen et al. 2011). The molars are characterized by deep, vertical wear grooves that suggest a possible but by no means certain fish component to the diet. When the skull and postcranial skeleton were described somewhat later (Thewissen et al. 2001), *Pakicetus* proved to have long, cursorial artiodactyl limbs with no evidence of aquatic adaptations. However, the structure of the ear region of the skull added confirmation that *Pakicetus* and related pakicetids are basal cetaceans. The tympanic bone was only loosely attached to the periotic bone, allowing it to vibrate independently, which gives a greater sensitivity to underwater sound (Nummela et al. 2004).

Protocetids

The next grade (Fig. 8.22c) is represented by *Ambulocetus* (Thewissen et al. 1994; Madar et al. 2002). It was 3–4 m in total length, and clearly adapted for at least a semi-aquatic, sealion-like life. The vertebral column is strongly built and the robust transverse processes on the caudal vertebrae indicate a muscular tail capable of dorso-ventral undulation contributing to axially driven locomotion, although probably lacking a horizontal cetacean fluke at this stage. The feet were enlarged, particularly the hind feet, and long, diverging digits indicate the existence of webbing, while the humerus and femur were shorter than in pakicetids. Functionally the limbs were paddles for aquatic locomotion but were still capable of supporting and moving the body on land in a pinniped mode of terrestrial locomotion. The skull of *Ambulocetus* has also acquired several new cetacean characters such as an elongated snout and shift of the external nostrils to a more posterior, dorsal position. The dentition is very whale-like, with incisors and canine forming parallel rows and the posterior teeth transversely compressed and high crowned, along with a reduced zygomatic arch, evidence for a mainly fish-eating habit.

Other taxa at about the same protocetid grade as *Ambulocetus* have been described. *Rodhocetus* (Fig. 8.22d) possessed relatively even longer hands

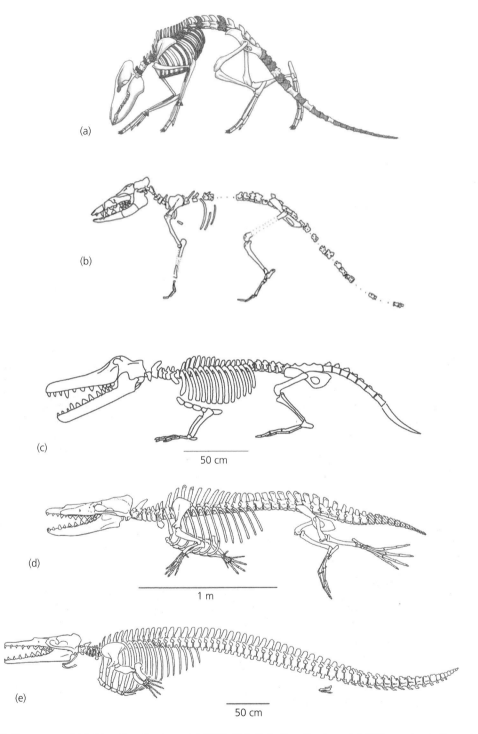

Figure 8.22 Stem ceaceans. (a) *Indohyus* (from Thewissen et al. 2007). (b) *Pakicetus* (from Thewissen et al. 2001, reproduced with permission from Nature Publishing Group). (c) *Ambulocetus* (from Thewissen et al. 2006). (d) *Rodhocetus* (from Gingerich et al. 2001, reproduced with permission from the American Association for the Advancement of Science). (e) *Dorudon* (from Gingerich and Uhen 1996).

and feet and shorter femur and humerus (Ginger-ich et al. 1994; Gingerich et al. 2001). The limbs of the 2–3 m long *Maiacetus* were relatively shorter, and one remarkable specimen is a skeleton of a fe-male with a near-term foetus in the body cavity. It is oriented for a head-first exit, which is typical of terrestrial mammals, and the reverse of modern whales (Gingerich et al. 2009).

Basilosaurids

The basilosaurids (Fig. 8.22e) such as *Dorudon* (Gingerich et al. 2009) represent a stage of cetacean evolution in which terrestrial locomotion was no longer possible, although small hind-limb elements were still present in addition to the paddle-like fore limbs. They were somewhat larger animals than the previous grades, with a body length around 3–4 m. The neck is shortened, the thoracic and lumbar vertebrae similar to one another, loss of contact be-tween the sacrum and the vertebral column, and a ball and socket tail vertebra indicating the presence of a horizontal fluke in life. The fore limb is modi-fied into a paddle, with flattened bones and limited mobility. The hind limb is extremely reduced al-though still recognizably artiodactyl in structure. The skull too shows several new cetacean features, notably an elongated, slender snout and the nos-trils shifted even further back. The basilosaurid ear is essentially like that of modern whales, for the bulla and tympanic bone are acoustically isolated from the rest of the skull by air-filled sinuses, indi-cating permanent underwater hearing (Nummela et al. 2004).

Modern whales, both the toothed odontocetes and the planktivorous mysticetes, constitute a monophyletic group, Neoceti. Ancestrally the den-tition was further reduced to a simple, homodont condition lacking any tooth occlusion, and the ear was further elaborated in minor ways. The postcra-nial skeleton, however, was little further evolved.

8.5.2 The pattern of acquisition of cetacean characters

The transition from a terrestrial to an obligator-ily aquatic habit, like the reverse, demands adap-tations in many aspects of the phenotype. In the case of the origin of whales, these include shifts to piscivory, aquatic locomotion, underwater hearing and vision, underwater parturition and suckling, plus physiological adaptations for prolonging times of complete submergence, thermoregulation, and the absence of a source of fresh water. The still quite limited fossil record of stem cetaceans offers some insights into these changes (Uhen 2007). There are indications that the initial step, as represented in pakicetid-grade taxa, was extension of feeding hab-its into water by a still terrestrial artiodactyl. Skel-etal adaptations were limited to relatively minor dental changes and an ear capable of a degree of subaquatic hearing. By the stage represented by protocetid fossils, the postcranial skeleton was sub-stantially modified for an amphibious mode of life akin to modern pinnipeds. The skull and dentition showed further specialization, specifically for fish eating as a significant part if not the whole of the diet. Underwater hearing was also enhanced and the nostril had shifted dorsally to enhance breathing while in the water. However, birth still occurred on land. The basilosaurid stage involved a shift to an obligatorily permanent aquatic, presumably marine life with caudal locomotion, fore-limb steering, and complete loss of hind-limb locomotory function.

With so few intermediate grades represented, little more can be said than that at each successive step evolutionary changes in body size, feeding (including dental structure and jaw shape), loco-motion (including axial, fore-limb, and hind-limb skeleton) and aquatic hearing all occurred, a pattern compatible with the correlated progression model of character evolution.

CHAPTER 9

A synthesis

9.1 The nature of palaeobiological explanation

Hypotheses about the causes of long-past events of which only incomplete traces remain constitute the historical as distinct from the experimental sciences. Of their nature, such events cannot be subjected to the kinds of direct observation to which the processes concerning a discipline like ecology can, let alone to the repeatable experimental methods available in, say, genetics. This difference from the conventional view of the scientific method provokes the question of whether historical hypotheses are properly part of science at all, since on the face of it they can never be directly corroborated or refuted. From time to time a hard line has been taken about supposed explanations for past evolutionary events, particularly those based on fossil evidence (e.g. Gee 2000). They have been rejected as no more than untestable scenarios, or even more pejoratively as 'just-so stories' with a nod to the delightful but biologically absurd explanations of Rudyard Kipling for the origin of various adaptations like the elephant's trunks and rhinoceros's skin. The brief influence of 'pattern' cladism in the 1980s was of such a cast (e.g. Platnick 1979, Nelson and Platnick 1984, and see Kemp 1985 for a contemporary review). It was claimed that the only valid scientific evidence available for discovering the natural classification of organisms is the empirically observed pattern of distribution of homologous characters amongst the taxa under investigation. Going further and stating, as most people did, that this pattern is a consequence of specifiable evolutionary changes and descent from common ancestors required assumptions about evolutionary events that are unobservable and therefore immune

to scientific testing: so representing the diversity of organisms as explicitly evolutionary trees is impermissible. Scenarios, which purport to explain such trees in terms of supposed processes like natural selection of particular adaptations, or external causes of extinctions and speciations, are regarded as even more in the realm of untestability. Thankfully this kind of extreme view of the worthlessness of attempts to account for particular evolutionary events that are recorded in the fossil record dissipated in the face of its inconsistency with generally accepted ideas of how science should be conducted (Kemp 1985, 1999), although there is still an echo of it to be found in the reluctance of many current palaeontologically oriented taxonomists to allow anything other than the simple, atomistic model of character evolution to determine hypotheses of phylogeny (see Chapter 3).

More recently, a number of philosophers of science have explored the methods of the historical sciences and the ways in which they differ from experimental approaches, and have concluded that perfectly logical historical scientific hypotheses can exist, hypotheses no less acceptable in principle than hypotheses based on experimental procedures (Kemp 1999; Cleland 2002; Jeffares 2010). Others remain a little more sceptical, notably Turner (2005), who was concerned about the irretrievable loss of information over time, and the impossibility of always distinguishing between alternative plausible explanations for the same empirical observations of the fossil record (Turner 2009). However, in the view of all these authors the fundamental point is that traces of past events associated with a particular evolutionary transition can be found today that throw light on the causes of the event, and that the greater the number of different kinds of

The Origin of Higher Taxa, First Edition. T. S. Kemp.
© T. S. Kemp 2016. Published 2016 by Oxford University Press and The University of Chicago Press.

traces found, the greater is the net constraint they impose on possible explanations of the event. The fossil record and the changes it records over time provides the most prominent clues, but all manner of geological and geochemical signals may be associated with the fossils and their environments that have a bearing on the event. Indeed, recent developments in palaeobiology have included several exciting new techniques and discoveries, from CT and synchroton scanning of fossils revealing finer and finer histological detail, to stable isotope and trace element evidence indicating aspects of the ecological circumstances of the life of a fossil taxon (Jeffares 2010; Kemp 2010). Armed with this array of evidence, a number of principles concerning how properly to use it need to be developed: how to combine the palaeobiological evidence of a particular event with neontological evidence of general mechanisms of evolution; how to handle missing and conflicting evidence; how to evaluate competing hypotheses that on the face of it can all give a plausible account of the event.

9.1.1 Combining the evidence

Deciphering the causes of unique historical events depends on combining observations of preserved historical traces of the event with knowledge of the general properties of the kinds of entities involved that have been derived from studies of modern examples. For example, the cosmological history of the universe derives from sightings of celestial bodies, energy measurements, etc., interpreted in the light of the known laws of physics; geology has long depended on the principle of uniformitarianism, whereby geophysical processes observably in action today are invoked as the causes, acting over long periods of time, of the geomorphological structures reflecting the earth's history. In the case of major evolutionary transitions inferred from the diversity of living and fossil organisms, the historical evidence available consists of their morphology along with whatever is determinable from the fossil record about their ages, geographical ranges, and the environmental conditions in which they occurred. In order to understand these traces of past life, it is necessary to know about the evolutionary properties and potentials of organisms in general.

This knowledge is derived from neontological studies of living organisms, such as the nature of phenotypic integration, the relationship between form and function, the rules of ecology and of natural selection, and the genetic processes capable of effecting evolutionary changes in characters. A hypothesis of the cause of the evolutionary event in question can be proposed on the basis of known evolutionary processes plausibly deemed to have brought it about.

This relationship between the evidence of particular events and the evidence of general processes is often referred to as the relationship between pattern and process. Here, 'pattern' refers to the nature of the organisms and their phenotypic characters; 'process' refers to the evolutionary mechanisms that brought about those patterns. In more general terminology, patterns are the observed *effects* and processes are the inferred *causes*, and the two are logically and inextricably combined in any account of a particular evolutionary event. A resulting explanatory hypothesis takes the form of a description of the historical course of the event coupled with the postulated evolutionary mechanisms that brought it about. As long as the perceived history is compatible with the proposed mechanism, the hypothesis is logically consistent.

9.1.2 Missing and conflicting evidence

However, the pattern–process issue is rarely so simple, especially in the case of major evolutionary transitions, because there are good grounds for expecting the fossil record to be highly incomplete. Stratigraphic incompleteness arises from the non-continuous formation of fossil-bearing strata at even very low temporal resolution. The vast majority of fossil taxa at lower taxonomic levels such as species and genera and often even family will have lived at a time and place not represented by sedimentary rocks, and therefore not represented by even a single specimen. It is true that there are occasional parts of the fossil record with a resolution as high as 10^4–10^5 years or even better, but these are extremely rare and invariably cover a relatively minute time span and geographical area. In the present context, even for a major transition taking tens of millions of years to run its course, representatives of all or nearly all the intermediate stages may be missing. Indeed, it is impossible to decide from the fossil record alone whether the transition did take so long, or whether it occurred very much more rapidly, even instantaneously. This incompleteness of palaeobiological information extends to ecological inferences because of the very limited indications in the rocks of the environmental circumstances and community relationships, even of those extinct taxa that are preserved as fossils.

A complete understanding of the evolutionary processes that bring about major changes would allow a good deal of compensation to be made for the incompleteness of the fossil record because it would be possible to predict with confidence what kinds of intermediate taxa and environmental conditions must have appertained during a transition. Unfortunately, no such full understanding exists because evolutionary mechanisms are studied in the modern, neontological context and are therefore restricted to the relatively very short term. Experimental and field investigations of natural selection are limited to no more than a very few phenotypic traits observed over at most a few decades, while the molecular basis of genetics and the effect of mutations on the development of the phenotype are largely restricted to a few generations. Very rare or slow-acting processes are unlikely to be revealed by such studies, yet in principle there is no reason why such processes should not exist and have a highly significant role in major evolutionary transitions, given the timescale involved. Furthermore, major transitions are complex events involving large numbers of phenotypic characters. If the fossil record was sufficiently complete, then it could indicate the existence of rare and slow processes by revealing patterns of taxonomic change not readily compatible with the known, ecological timescale mechanisms of evolution. If, for example, the existing fossil record of the origin of new invertebrate phyla in the Cambrian was somehow known to be virtually complete, it would demonstrate the inadequacy of the familiar evolutionary process of gradual, adaptive change to explain this particular case. Alternative processes would need to be postulated.

The default assumption of mainstream evolutionary biology is that all evolutionary change at whatever taxonomic level can be explained by known

ecological timescale or microevolutionary processes acting for long enough time. Mainly these include natural selection of single or small numbers of alleles in a continuously interbreeding population, but also the effects of random genetic drift, immigration/emigration, and genetic heterogeneity on the gene-pool structure. The fundamental justification of this extrapolationist view is that if the event can be explained adequately by known processes, then it is less parsimonious to invoke unknown mechanisms. This simple statement is unexceptionable save in one respect. By accepting it uncritically, the possibility of discovering possible slow-acting, long-term processes is excluded, even in principle. The complexity of changes involved in the origin of a higher taxon, coupled with the relative dearth of actual evidence about its course, means that an account can always be forced into the straightjacket of microevolutionary processes. Suitable missing intermediate stages can be invoked and appropriate environmental circumstances imagined that complete the simple adaptive scenario. But what if there actually were slow-acting, long-term, processes at work? Are they destined to remain undiscovered? If the fossil record is to contribute anything of importance to evolutionary theory, it is revealing what happens over very long periods of time and whether there are indeed additional processes on this timescale invisible in ecological time.

Pattern and process are logically related to one another as effects and their causes, and therefore epistemologically they feed off one another. Observations of the fossil record offer insights into the kinds of processes that could have caused the evolutionary event. Simultaneously, neontological studies of organismal biology, genetics, and ecology offer insights about the kinds of intermediate organisms and patterns of character changes expected to exist in the course of the event. If what is known from the fossil record is entirely compatible with what has been discovered about the evolutionary properties of living organisms in general, then the matter is straightforward enough because explanations of the former entirely in terms of the latter are possible. Frequently however, no such compatibility exists, as for example in the origin of most higher taxa, where the fossil record taken at face value indicates at best a sequence of instantaneous, quite large evolutionary transitions, even though current evolutionary theory requires them in reality to have been gradual. This conflict of the evidence arises either because most or all of the intermediate stages of the transition are not preserved, or because knowledge of mechanisms of major evolutionary change are not adequately understood. Few evolutionary biologists or philosophers have addressed this issue (Grene 1958; Kemp 1988; Kemp 1999). Historically, priority has usually been given to either the process or the pattern depending on the background and proclivities of the investigator, rather than an attempt made to reconcile the two (Fig. 9.1a–c). A somewhat literal reading of the fossil record led to such ideas as Henry Fairfield Osborn's concept of aristogenesis as an internal driving force, and Otto Schindewolf's typostrophism as a means of instant origin of new kinds of higher taxa: in effect they assumed that there were as yet unknown evolutionary processes. Conversely, George Gaylord Simpson and other architects of the Modern Synthesis interpreted the fossil record in terms of the neo-Darwinian mechanism of gradual, adaptive change in continuous interbreeding populations, with the implication that the fossil record is extremely incomplete because of the absence of large numbers of intermediate stages showing a pattern of gradual transition: in effect they assumed the existence of as yet undiscovered fossils. In a perceptive essay on the contribution of palaeobiology to evolutionary theory, Gould (1983) described the first of these as the 'irrelevant' phase and the second as the 'submissive' phase of palaeobiology. The third, current phase he described as a 'partnership'

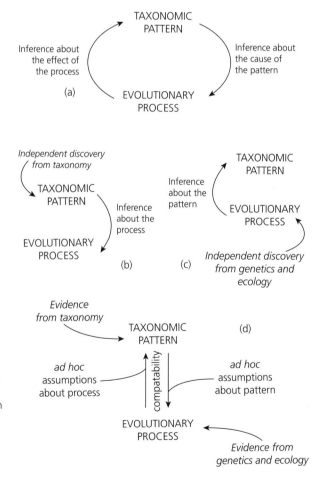

Figure 9.1 The epistemological relationship between taxonomic pattern as observed evolutionary outcomes and evolutionary processes as revealed by experiment and observation of living organisms. (a) The apparent circularity of using observed pattern to infer evolutionary processes while at the same time using the inferred processes to infer the correct pattern. (b) The option of giving the observed taxonomic pattern hegemony. (c) The option of giving the discovered evolutionary processes hegemony. (d) Evolutionary theory seen as the creation of compatibility between incompletely known patterns with incompletely known processes by creating ad hoc assumptions about either or both. See text for further explanation. (From Kemp 1989.)

in so far as palaeobiology does have a valid input into evolutionary theory, although it was not entirely clear exactly what this means.

Kemp (1999) suggested how to avoid what amounts to a logically arbitrary choice being made when either the perceived pattern (Fig. 9.1b) or the known process (Fig. 9.1c) is assumed to be paramount. Pattern and process should be treated for what they are: respectively, the effect and cause of that effect (Fig. 9.1d). An explanatory hypothesis requires the input of both, and that the two are compatible with one another. The perceived pattern of taxa and characters in the fossil record must be such as to be capable of being generated by the proposed evolutionary cause; the proposed evolutionary cause must be such as to be able to generate the perceived pattern. The question is then how to achieve compatibility when there is incongruence between the empirically observed fossil record and the received evolutionary theory. The answer is simple in principle. Additional assumptions about the pattern and/or about the processes at work have to be made. These are technically ad hoc assumptions because they are neither corroborated nor refuted but are taken as true in order for the hypothesis to be true. As has already been made clear, by far the commonest and least controversial ad hoc assumption is that the fossil record is incomplete, and that all those intermediate stages required of a neodarwinian interpretation are missing due to lack of preservation. However, it is also possible and no less principled to make assumptions about unknown evolutionary processes, and examples of this have arisen in earlier chapters. For example, there are several cases in the fossil record and in comparative anatomy described in Chapter 5 that are most easily explained as heterochronic modifications during development, despite an almost complete lack of knowledge from modern genetic studies of how mutations might bring it about. The ideas discussed in Chapter 6 about ecological drivers of major evolutionary transition are similarly based more on inferences from the nature of organisms than on empirical studies of long-term, multitrait selection in modern environments. A good deal of recent molecular developmental genetics has revealed unexpected, often surprising phenomena such as hierarchically organized genetic developmental networks, conserved regulator gene

families, and so on, pointing to possible, hitherto unknown evolutionary mechanisms (see Chapter 5, section 5.3.2). Relating these to the patterns of evolution in the fossil record is in its infancy, but will doubtless prove fundamental in explaining major evolutionary events.

9.1.3 Evaluating palaeobiological hypotheses

Of course, appreciating this underlying structure of explanatory hypotheses in palaeobiology scarcely solves much as it stands because there will always be any number of ad hoc assumptions about missing fossils and/or unknown mechanisms that could be made, and therefore a number of hypotheses are possible all of which could account for a particular evolutionary transition. A method is required for selecting which one amongst them is the most acceptable. In keeping with the standard scientific principle of simplicity, or Occam's Razor, the preferred hypothesis is the one with the least ad hoc burden. The less uncorroborated auxiliary assumptions made, the simpler is the explanation. Another way to put it is that the less the ad hoc burden, the less the capacity for the hypothesis to be wrong because one of the ad hoc assumptions is wrong. To mention the 'Just-so stories' again, the reason logically why Kipling's explanation of the origin of the elephant's trunk is rejected is not because it is *known* to be untrue—no-one was present at the alleged incident making empirical observations—but because it demands an absurdly large amount of ad hoc assumptions, such as that a baby elephant's nose could be elongated by being pulled, that crocodiles could hang on for long enough to cause this, that the ecophenotypic effect of nose lengthening could be inherited. No biological processes for any of these has been revealed by modern studies of elephants, crocodiles, or anything else.

The principle of minimal ad hoc assumptions works well and is largely taken for granted for simple systems, such as statistical methods for drawing best-fit graphs of scattered points in an experiment. In the case of complex evolutionary events involving many variables and a diverse array of different kinds of incomplete evidence, no such easy quantitative evaluation of hypotheses is possible. How many missing fossils equals how much ignorance about genetic mechanisms, or lack of evidence of palaeoecological conditions? For example, in the case of the Cambrian explosion, selecting a coherent hypothesis of its cause requires a choice amongst numerous ad hoc assumptions: a long sequence of missing intermediate stages in the fossil record; unreliability of the molecular clock; the importance of oxygen levels for rate of evolution of large body size; the selection effect on a community of biotic innovation such as predation; the role of calcium balance in the early Cambrian seas; a genetic developmental mechanism that responded by generating a wide range of short-lived lineages. The selection made amongst all these and other assumptions depends more on the proclivities of particular authors than any attempt to estimate the size of the ad hoc burdens of different accounts. At least it is possible from this point of view to make clear just what the ad hoc components and what the empirically corroborated components of a hypothesis are.

9.1.4 It may be true but is it science?

In this somewhat rarefied light of the epistemology of palaeobiology, how much confidence can be expressed in hypotheses about the causes of the origin of specific higher taxa? In previous chapters, evidence has been explored from the fossil record and its palaeoecological settings, and from modern organism biology, developmental biology, and ecology. In any given case of a major transition, can all these categories of information be combined, made compatible by ad hoc assumptions where necessary, and moulded into a single account of the causes of the event that has scientific validity? Views will differ along a spectrum of scepticism. At one extreme, such hypotheses can be regarded as in perfectly acceptable scientific form. They are explanations of real, rational events, and they are open to potential testing by discovering new fossils or new information about, say, molecular developmental genetics or physiological functioning in analogous living organisms. Computer modelling of the general behaviour of complex integrated systems may also add to or detract from the level of support for the hypothesis. A more sceptical view is that such hypotheses are indeed in proper scientific form, but with so many variables and so little information available they are not realistically testable. Therefore

they are no more than plausible, which is to say explanations that could be true in the light of the very limited knowledge about the event, but in which very little confidence can be placed. The most sceptical view is that the level of testability is so low as to render them worthless as scientifically acceptable hypotheses. From this perspective, the accounts are not hypotheses at all, but at best only descriptive devices to accommodate a limited amount of knowledge in a coherent picture—scenarios in a pejorative sense. To hypothesize, for example, that the origin of tetrapods was caused by natural selection for a specified range of ecological parameters acting on a taxon of shallow-water fish exposed to an ecological gradient that existed between water and land in the Late Devonian (Chapter 6, section 6.3) actually conveys no further understanding than that certain morphological characters differed via two or three known intermediate grades between *Eusthenopteron* and *Acanthostega*, and that during the period when these animals existed, certain geochemically revealed environmental conditions apertained. The rest of the scenario is just an appealing framework on which to hang these observations with no further predictive meaning.

Perhaps the best way to view the situation is that these three different levels of evaluation—testable hypothess, plausible picture, and imaginative scenario—apply to different respective cases. The origin of mammals, and perhaps of birds is in the first category because so much is known about intermediate-grade fossils, the palaeoecological conditions, and the biological significance of the characters. The origin of tetrapods, turtles and whales, and perhaps optimistically of molluscs, echinoderms, and arthropods fall into the second category. The origin of other invertebrate phyla and classes, where so very little is known about either intermediate stages or the environmental circumstances, are scarcely beyond speculation yet.

It can be concluded that the investigation of the causes of the origin of higher taxa is certainly a valid scientific endeavour in so far as it is accepted that higher taxa did in fact arise as real events in the history of life. It also appears to be a worthwhile endeavour in so far as explanatory hypotheses in some cases are based on sufficiently robust, wide-ranging evidence to be acceptable, even if others fail any test but plausibility.

The next question is whether a general picture of the causes of the origin of higher taxa can be distilled from the evidence in particular cases.

9.2 A summary of the evidence

The evidence reviewed extensively in earlier chapters may be summarized briefly before attempting to bring it together in a general view of the origin of higher taxa. To an approximation, the kinds of evidence available about how higher taxa originate falls into the two categories described. Pattern evidence is about the outcome of evolutionary change and consists of the morphology of fossils determined by cladistic analysis to be of intermediate grades, along with geomorphological and geochemical traces of the environments in which they occurred. It also embraces evidence from living representatives of the higher taxon, if any, about how the end product is constructed and functions as an integrated organism. In some cases, related living taxa can also offer important if indirect information about the nature of ancestral phenotypic stages. The principal limitation to pattern evidence is the incompleteness of the fossil record, aggravated by the lack of actual ancestral taxa for study.

Process evidence is derived from genetic and ecological studies that have revealed the known evolutionary processes that were available to all the organisms, and the nature of constraints and controls on the directions of evolutionary trajectories that may have prevailed. The principal limitation of process evidence is the extremely brief length of time over which evolutionary processes can be observed in living organisms. Most studies in both field and laboratory of such processes as gene mutations and their effects, maintenance of phenotypic integration by developmental feedback and pleiotropy, and natural selection take place over at best a matter of a few decades such as Lenski's *in vitro E.coli* studies (Pennisi 2013).

9.2.1 Evidence from the nature of living organisms

Centuries of study of the biology of whole organisms has revealed two generalizations that impose constraints on possible causes of major evolutionary change. The first is that taxa consist of discrete clusters of organisms in morphospace that are significantly separated from other such clusters, and higher taxa are no exception. Therefore the extinction of the species and multispecies taxa that constitute a diverging lineage is as fundamental a part of the process as the origin of new intermediate species and multispecies taxa. This has implications for what the relationship is between the lineage's trajectory through morphospace and the environment within which it is evolving.

The second fundamental property of organisms is the structural and functional integration amongst the arbitrarily defined characters. They are systems of correlated parts rather than collections of atomistic traits. The implication of this, again commonplace observation is that evolving lineages must consist of sequences of intermediate stages each of which is a fully integrated phenotype. At the molecular level, genetic mutations affecting the phenotype and therefore available for selection tend to appear singly and uncorrelated with other mutations, while natural selection acts on overall fitness which results from the integrated effect of all the genes at once. This raises the evolvability question of how the organism can remain functionally integrated while at the same time changing gene by gene and trait by trait over long distances through morphospace. There is evidence for the existence of three possible mechanisms. Developmental feedback consists of mechanisms for maintaining integration by direct compensatory interactions between functionally related parts during their development, of which there are mechanisms of several kinds. Modularity of construction potentially generates evolvability by the relative independence of modules from one another, so that any one module is more or less free to evolve without constraints imposed by having to remain tightly integrated with the rest of the organism. However, comparative evidence indicates that modules at best are relatively short-term, transient entities and therefore can only act as independent evolutionary units for at most brief periods of evolutionary time. Correlated progression refers to a pattern of successive small changes in functionally linked traits, hand-in-hand as it were, allowing over time long treks through morphospace culminating in large changes in many characters, but at no point losing integration amongst them all. The driving force is natural selection acting on the overall fitness of the organism. At the level of higher taxa, it was concluded that correlated progression is the most important mechanism for maintaining integration while permitting major evolutionary transition.

9.2.2 Evidence from computer simulations of the evolution of complex systems

There are various ways in which computer simulations of the properties and behaviour of complex systems can be investigated. Studies relevant to the origin of higher taxa include some that offer support for the idea that a correlated progression of single small steps in numerous functionally integrated characters can lead to an overall improvement in some defined measure of performance of the system. Other studies illustrate the effect of simultaneous selection on a number of characters, and demonstrate that the overall fitness of the modelled organism can increase by a range of different pathways, resulting from different patterns of random change in individual characters. These results indicate that in the complex world of natural selection acting on real organisms, where many characters are genetically variable and all potentially contribute to fitness as an emergent property, there is a random element to the actual direction through morphospace followed by an evolving lineage.

9.2.3 Evidence from the fossil record

The small number of cases in which the fossil record provides several intermediate grades of phenotype between an ancestral grade and a representative of a derived new higher taxon illustrate something of the pattern of acquisition of derived characters. The best case by far is that of the origin of mammals from the last common ancestor they shared with modern reptiles plus birds, and here the pattern supports the correlated progression model. Other cases amongst the vertebrate fossil record, notably those of tetrapods, birds, whales, and turtles, are also compatible with this general pattern although they offer less evidence. Accepting the arbitrary nature of the amount of phenotypic change

required in order to announce the origin of a new higher taxon, these cases also give a sense of the length of time it actually takes to assemble evolutionary changes to this point, namely of the order of tens of millions of years.

The origins of the invertebrate higher taxa are very poorly represented by fossil evidence of intermediate grades, notwithstanding rather optimistic claims for several Cambrian forms. There is too little evidence to decide whether the acquisition of the derived characters of the crown phyla corresponds most closely to a correlated progression pattern, a modular pattern, or for that matter a virtually instantaneous revolutionary pattern. Furthermore, the fossils are not very helpful in illustrating the rate of acquisition of the derived characters of crown phyla, since the latter appear more or less fully evolved in the Cambrian, with virtually no prior record. Whether it was a case of very rapid evolution, as a literal reading of the fossil record indicates, or more in keeping with the molecular evidence of a much earlier diversification into the incipient phyla remains a matter of controversy.

The fossil record also illustrates how an evolving lineage which culminates in a new higher taxon is far from a monotonic sequence of species-level taxa. At points along its progression through morphospace, radiations into several or sometimes many lower level taxa occur, a phenomenon that has implications for the nature of the driving force of the lineage.

9.2.4 Evidence from developmental biology

Embryonic evidence offers an indirect, potentially ambiguous, and frequently misleading window into ancestral stages of living organisms and the pattern of acquisition of derived characters of a higher taxon. However, it does reveal a number of general processes that can bring about major transitions whilst maintaining phenotypic integration. These include heterochrony and heterotopy, in which the relative sizes and positions respectively of different parts can alter. There is a surprisingly large number of specific cases in which heterochrony seems to have been significant. Allometry, in which the relative sizes of parts differs with overall body size, also seems to have been a significant evolutionary process, especially in association with miniaturization.

These process are inferred from comparative anatomy of embryonic stages and very little is yet known about how integration amongst the parts is maintained, not even in most cases whether it is a result of interaction at the molecular level, or the result of natural selection of the best combinations amongst independently variable traits. What is known is that mechanisms of developmental feedback are important generally in development. The direct effects of cell types and developing tissues on one another in the embryo can ensure functional integration of the different tissues that constitute an organ such as a limb. The implication is that a mutation causing an evolutionary modification of one such tissue will automatically lead to appropriate adjustments in others so that the organ continues to function coherently. At the molecular level there are several mechanisms for this, particularly a variety of types of cell-to-cell signalling systems.

The regulation of development at a molecular level is based on genetic regulatory networks, which are quasi-hierarchically organized systems of modules consisting of sub-networks of genes and gene products. In outline, at the highest level are highly conserved modules which determine the position in the embryo where a structure is to develop. A battery of somewhat less conserved modules of various sorts regulate the activity of the genes that transcribe the functional protein and RNA molecules directly controlling the activities of the differentiating cells. This organization is of the highest theoretical significance for understanding major evolutionary transformation and, although insufficient detail is yet known to be able to account comprehensively for any particular case, a number of principles are understood. One is the role of cis-regulation, in which several to many genes code for transcription factors affecting a single target gene. This permits a mutation in a regulator gene to affect the activity of the target gene, such as the time or location of its expression, by only a small amount and so it may remain adequately integrated with the activities of the rest of the genome. Thus, the regulatory genome can evolve in a correlated progressive fashion analogous to the evolution of phenotypic characters. Many successive small affect mutations

can lead to large evolutionary transitions without loss at any stage of adequate integration of the whole complex system. A second general principle that has emerged is that conserved modules, especially perhaps those in the middle of the hierarchy often referred to as 'plug-ins', can evolve to be deployed in new regions of the embryo, and regulate the development of new traits. A given module is likely to be involved in a wide variety of different structures; a given structure invariably requires the activity of a wide variety of modules for its development. Again, the process of cis-regulation can allow this to occur without initially causing large changes that would be likely to be inviable.

9.2.5 Evidence from ecology

Empirical evidence of the environmental and community circumstances of major evolutionary transitions is very scant, and therefore the ecological perspective on the origin of higher taxa is mainly derived from predictions based on the correlated progression model of character change. From it comes the implication that natural selection acts on overall fitness that is contributed to by all the phenotypic traits together. The effect is that many different trait combinations may endow the organisms with a similar level of overall fitness, and so the extent of neutral drift in phenotypic traits is likely to be significant. The model also predicts that in principle all the parameters of the environment contribute to the overall selection pressure experienced by the evolving lineage, each parameter associated with one or more different functions or structures of the organism. Thus the lineage can be visualized as traversing a highly multidimensional adaptive landscape. The directionality of the evolving lineage leading to a new higher taxon, in which many traits have evolved in a broadly trend-like fashion to reach very different values, implies that the ecological parameters to which they were responding varied in a gradual way, and together they constituted a compound ecological gradient. This is what the lineage tracked over evolutionary time, and in terms of the adaptive landscape, it can be described as following a ridge representing the multiple graded parameters. Over time, new adaptive space is continually entered by following the ecological gradient, rather than seeing it as fitness continually increasing in a constant part of the adaptive space. The adaptive ridge must be modelled as more or less flat-topped as well as horizontal in order to visualize low taxonomic level radiations occurring at different places along the ridge, as represented by the short-lived diversifications of intermediate grades of evolution that occur along the lineage.

There is some evidence from the palaeoenvironmental observations that general ridges of this sort do exist that are correlated with major lineages. Notably, the geochemical evidence for rises in oxygen and calcium levels in the Late Proterozoic and Cambrian suggest this, as does the simultaneous evolution of increasingly complex communities of increasingly specialized taxa. In the case of the origin of tetrapods, the gradual increase in the population density and the size of land plants, and of the increasing availability of land-based invertebrate food are suggestive of ecological gradients. For the endothermic higher taxa, birds and mammals, the number of correlated changes related to a variety of environmental parameters such as ambient temperature range and the degree of seasonal aridity can best be interpreted as evidence for the existence of a compound ecological gradient having driven the lineage.

9.3 Conclusion: a general picture of the origin of higher taxa

My aim in writing this book has been to investigate our current understanding of how higher taxa arise, which is to ask how and why an evolving lineage of ancestral and descendant phenotypes should trek such a long distance through morphospace as to culminate in what we can agree is a new higher taxon. I accept the arbitrary nature of the category 'higher taxon', but since the treks themselves are real enough, what we choose to label the end-product is not very important. Modern thinking in biology is dominated by the concept of systems of interrelated parts, rather than by the more atomistic view of the past which tended to see organisms in terms of discrete characters, genomes in terms of strings of genes each with a more or less fixed fitness value,

and niches and habitats as lists of biotic and abiotic parameters. Unquestionably, the atomistic approach successfully organized a vast amount of biological knowledge, but we must acknowledge that the systems view is a far better representation of the reality of biological phenomena, notwithstanding what are still formidable epistemological barriers to their analysis. Regarding our present quest, there are three relevant and interacting systems. First is the organism, which is a system of functionally integrated structures and physiological processes; the second is the genome, which is a regulatory system of genes and their products interacting both with one another and with environmental cues; the third is the environment, which is a system of interacting physical parameters, chemical quantities, and species in a biotic community. These three systems are intimately related to one another in interesting and complex ways. The genome generates the phenotype but is itself altered by natural selection acting on the phenotype; the environment provides the selection force on the phenotype, but aspects of the environment are themselves determined by phenotypic activities, which may be physical such as gas exchange, chemical such as energy flow, and biotic such as community structure. And since therefore the environment affects the phenotype and the phenotype affects the genome, then the first and the last are themselves interconnected. Major evolutionary transformation, in common with evolution at every level, is a process involving all three systems, and no interpretation of it could be complete without taking all into account. This, of course, explains why evolution at this level it is such a neglected field compared to the vastly simpler, more tractable case of short-term microevolution.

I have held in mind throughout the two fundamental questions that need answering: what mechanisms maintain a high level of phenotypic integration while simultaneously permitting extensive evolutionary change; and what drives and controls the direction of the evolving lineage as it treks its long distance through morphospace? Naturally, each particular specific case looked at has its own unique contingent circumstances, and so no single hypothesis or simple model is likely to explain them all. However, several general processes potentially do apply universally at this evolutionary level.

Three categories of mechanisms for the maintenance of phenetic integration during the course of extensive evolutionary transition exist. During development, feedback processes amongst individual tissues and parts of structures may ensure that a change in one is accompanied automatically by compensatory adjustments in others so that the ensuing structure remains integrated. The mechanism may be located at the level of direct cell-to-cell interactions, or mediated by underlying interaction amongst regulatory genes affecting the developmental pathways of the different constituents. Developmental feedback is certainly important, indeed more or less universal as a means of maintaining structural and functional coherence in the face of relatively minor genetic or environmentally induced perturbations, and is therefore of significance over short-term, low taxonomic-level evolution such as modifications to the proportions of limbs in related species. However, it is necessarily limited to tissues and parts of the organism that are contiguous with one another and so cannot be responsible for integration amongst separate anatomical structures and physiological processes; for example, the complex of different parts associated with temperature physiology or feeding strategy.

The second category is modularity, in which parts of the phenotype behave as semi-independent modules that can evolve in relative isolation of the rest of the organism. Despite the popularity of modularity as an explanation of evolvability, its role too is necessarily limited to relatively short-term evolution. In principle this is because much of the phenotype is not modular but consists of functional processes integrated with the rest of the organism. Empirically, modules are demonstrably transient, with their components changing over evolutionary time, so they cannot be long-term evolutionary units, and there is no satisfactory genetic basis to their putative individuality. Nor does the fossil record offer support for long-term evolution facilitated by modularity.

The third category is correlated progression, the mechanism of change in which small modifications to single parts of the phenotype are acceptable because there is enough functional flexibility between them to prevent loss of adequate integration and therefore of fitness. But no one part can change by

more than a small amount unless and until appropriate compensatory changes in the functionally linked parts have accumulated over evolutionary time. The process is driven by natural selection acting on the emergent fitness of the organism as a whole, which is a property compounded from all its traits. At any one moment in evolutionary time, some traits have stronger functional linkages with certain other parts, and these groups of traits will briefly have modular properties. Subsequently however, stronger linkages will come to be associated with different sets of traits, in an ever-changing pattern. Unlike either developmental feedback or modularity, correlated progression as a process sets no limits to how much evolution can occur in a lineage, that is so say how far through morphospace it can travel. It is therefore the principle mechanism responsible for the origin of major new taxa, during which large changes in many traits accumulate. The genetic basis of correlated progression resides in the architecture of the genetic regulatory system, although little is yet known in detail about this. The broadly hierarchical organization of the genetic regulatory system permits the occurrence of mutations having early-acting effects, which are necessary to account for the large differences that distinguish higher tax from one another. Crucially, however, because of the dampening effect of cis-regulation of gene expression by upstream regulators, such mutations initially have relatively small effects that do not significantly reduce phenotypic integration and therefore fitness. Over time, early-acting mutations can accumulate, with an increasing marked effect, species by species, culminating eventually in the large differences characterizing a new higher taxon.

The question of what drives a lineage in the direction it takes towards what is to become a new higher taxon is at present more speculatively answered. Given the propensity for very large change to result from the accumulation over time of small, single mutations by correlated progression, there is no reason to argue for any kind of large, macro-mutational steps. Indeed, there are good reasons to exclude that possibility, namely the logical construction of organisms, and the absence of empirical evidence from palaeontological or neontological study that it occurs. Generation by generation and species by species, the genetic variation of the respective phenotypes constituting the lineage can be small, and subject to natural selection.

However, natural selection acts on the phenotype as a whole, and the majority of the traits are subject to heritable variation and therefore contribute to overall fitness. Selection of the overall fittest therefore involves selection of the best compromise amongst the values of many traits, never just of a single one or even a few. With so many individual traits varying at the same time, some trait values will accumulate by chance because they have a neutral, or even slightly deleterious effect on the phenotype, due only to their occurrence as part of the overall fittest phenotype. Nevertheless, the main director of the evolving lineage is natural selection. The corollary of having many variable traits contributing to the overall selection pressure is that the selection pressure is compounded of the many parameters of the environment making simultaneous adaptive demands of all the structures and functions of the organism. This can be expressed as a high-dimensional adaptive landscape in which each dimension is an environmental parameter that varies as an ecological gradient, for example increasing temperature, reducing humidity, increasing prey size, and increasing predator efficiency and so on. It is this form of compound gradient that a lineage culminating in a new higher taxon must have tracked over evolutionary time. Metaphorically, the lineage moves along an adaptive ridge as it adapts to changing values of all the parameters. The process is certainly not constant in rate of evolution, and there are periods when the extension of the ridge has not yet arisen. The ridge also usually disappears behind the lineage, so to speak, as environmental conditions change including the change caused by the evolution of the lineage itself. There are periods when local, lower taxonomic-level radiations occur at regions along the ridge, as represented by diversification of intermediate grades of phenotypes. The kind of multidimensional ridge envisaged includes such major ecological transitions as the geochemical revolution at the start of the Cambrian, miniaturized life, free-living life to parasitism, water to land and vice versa, and low-energy ectothermy to high-energy endothermy. They are relative uncommon during the evolution of life on earth, which is why new higher taxa arise but rarely, compared to the frequency of lower taxonomic level transitions.

I have tried to show that of the cases of the origin of higher taxa that we know something about, especially from the fossil record, all are compatible with and can be explained by these processes. None point positively to any other processes such as macromutation, or particularly unrealistic environmental scenarios. None therefore require an explanation in terms other than existing hypotheses of how evolution happens, although a neo-Darwinian model based on natural selection of single characters in response to single, simple selection forces is not on its own adequate.

References

Aguilée, R., et al. (2012). Adaptive radiation driven by the interplay of eco-evolutionary and landscape dynamics. *Evolution* **67**: 1291–306.

Aguinaldo, A. A., et al. (1997). Evidence for a clade of nematodes, arthropods and other moulting animals. *Nature* **387**: 489–93.

Ahlberg, P. A. (1995). *Elginerpeton pancheni* and the earliest tetrapod clade. *Nature* **373**: 420–5.

Ahlberg, P. A., et al. (2005). The axial skeleton of the Devonian tetrapod *Ichthyostega*. *Nature* **437**: 137–40.

Ahlberg, P. E. (2011). Humeral homology and the origin of the tetrapod elbow: a reinterpretation of the enigmatic specimens ANSP 21350 and GSM 104563. *Palaeontology* **86**: 17–29.

Ahlberg, P. E., et al. (1996). Rapid braincase evolution between *Panderichthys* and the earliest tetrapods. *Nature* **381**: 61–4.

Ahlberg, P. E., et al. (2008). *Ventastega curonica* and the origin of tetrapod morphology. *Nature* **435**: 1199–204.

Aldridge, R. J., et al. (1986). The affinities of conodonts—new evidence from the Carboniferous of Edinburgh. *Lethaia* **19**: 279–91.

Aldridge, R. J., et al. (1993). The anatomy of conodonts. *Philosophical Transactions of the Royal Society B* **340**: 405–21.

Aldridge, R. J., et al. (2007). The systematics and phylogenetic relationships of vetulicolians. *Palaeontology* **50**(1): 131–68.

Alexander, D. E., et al. (2010). Model tests of gliding with different hindwing configurations in the four-winged dromaeosaurid *Microraptor gui*. *Proceedings of the National Academy of Sciences* **107**(7): 2972–76.

Algeo, T. J., et al. (2001). Effects of the Middle to Late Devonian spread of vascular land plants on weathering regimes, marine biotas, and global climate. In P. G. Gensel and D. Edwards (eds), *Plants Invade the Land—Evolutionary and Environmental Perspectives*. New York, NY: Columbia University Press, 213–36.

Allin, E. F. (1975). The evolution of the mammalian middle ear. *Journal of Morphology* **147**: 403–38.

Alonso, P. D., et al. (2004). The avian nature of the brain and inner ear of *Archaeopteryx*. *Nature* **430**: 666–9.

Anderson, P. S., et al. (2013). Late to the table: diversification of tetrapod mandibular biomechanics lagged behind the evolution of terrestriality. *Integrative and Comparative Biology* **53**(2): 197–208.

Andrews, S. M. and T. S. Westoll (1970). The post-cranial skeleton of *Eusthenopteron foordi* Whiteaves. *Transactions of the Royal Society of Edinburgh* **68**: 207–29.

Arnold, S. J., et al. (2001). The adaptive landscape as a conceptual bridge between micro- and macroevolution. *Genetica* **112–113**: 9–32.

Arthur, W. (1997). *The Origin of Animal Body Plans: A Study in Evolutionary Developmental Biology*. Cambridge: Cambridge University Press.

Atchley, W. R. and B. K. Hall (1991). A model for development and evolution of complex morphological structures. *Biological Reviews* **66**: 101–57.

Ax, P. (1987). *The Phylogenetic System*. Chichester: John Wiley.

Baer, K. E. von (1828). *Entwicklungsgeschichte der Thiere: beobachtung und Reflexion*. Königsberg: Erster Theil, mit 3 col. Kupfertaf.

Baier, D. B., et al. (2007). A critical ligamentous mechanism in the evolution of avian flight. *Nature* **445**: 307–10.

Bakker, R. T. (1972). Anatomical and ecological evidence on endothermy in dinosaurs. *Nature* **238**: 81–5.

Barton, N. H. and L. Partridge (2000). Limits to natural selection. *BioEssays* **22**: 1075–84.

Bengtson, S. (2004). Early Skeletal Fossils. *Paleontological Society Papers* **10**: 67–77.

Bengtson, S., et al. (2012). A merciful death for the 'earliest bilaterian' *Vernanimalcula*. *Evolution and Development* **14**(5): 421–7.

Benton, M. B. (2014). How birds became birds. *Science* **345**: 508–9.

Benton, M. J. (2000). Stems, nodes, crown clades, and rank-free lists: is Linnaeus dead? *Biological Reviews* **75**: 633–48.

Benton, M. J. (2004). *Vertebrate Palaeontology*, 3rd edition. Oxford: Blackwell Scientific.

Benton, M. J. and P. C. J. Donoghue (2007). Paleontological evidence to date the tree of life. *Molecular Biology and Evolution* **24**(1): 26–53.

Berner, R. A. (2009). Phanerozoic atmospheric oxygen: new results using the GEOCARBOSULF model. *American Journal of Science* 309: 603–6.

Berner, R. A., et al. (2007). Oxygen and evolution. *Science* 316: 557–8.

Blair, J. E. and S. B. Hedges (2005). Molecular phylogeny and divergence times of deuterostome animals. *Molecular Biology and Evolution* 22(11): 2275–84.

Boisvert, C. A. (2005). The pelvic fin and girdle of *Panderichthys* and the origin of tetrapod locomotion. *Nature* 438: 1145–7.

Boisvert, C. A. (2009). The humerus of *Panderichthys* in three dimensions and its significance in the context of the fish-tetrapod transition. *Acta Zoologica* 90 (Supplement 1): 297–305.

Boisvert, C. A., et al. (2008). The pectoral fin of *Panderichthys* and the origin of digits. *Nature* 456: 636–8.

Bolker, J. A. (2000). Modularity in development and why it matters to evo–devo. *American Zoologist* 40: 770–6.

Bonaparte, J. F., et al. (2003). The sister group of mammals: small cynodonts from the Late Triassic of Brazil. *Revista de Paleontologia* 5: 5–27.

Borycki, A.-G. (2004). Sonic hedgehog and Wnt signaling pathways during development and evolution. In G. Schlosser and G. P. Wagner (eds), *Modularity in Development and Evolution*. Chicago, IL: Chicago University Press, 101–31.

Botha, J., et al. (2005). The palaeoecology of the non-mammalian cynodonts *Diademodon* and *Cynognathus* from the Karoo Basin of South Africa, using stable light isotope analysis. *Palaeogeography, Palaeoclimatology, Palaeoecology* 223: 303–16.

Bourlat, S. J., et al. (2008). Testing the new animal phylogeny: a phylum-level molecular analysis. *Molecular Phylogenetics and Evolution* 49: 23–31.

Brazeau, M. D. and P. E. Ahlberg (2006). Tetrapod-like middle ear architecture in a Devonian fish. *Nature* 439: 318–21.

Briggs, D. E. G., et al. (1983). The conodont animal. *Lethaia* 16: 1–14.

Brink, A. S. (1965). A new ictidosuchid (Scaloposauria) from the *Lystrosaurus*-zone. *Palaeontologia Africana* 9: 129–38.

Bromham, L. (2003). Molecular clocks and explosive radiations. *Journal of Molecular Evolution* 57: S13–S20.

Brusatte, S. L., et al. (2010). The higher level phylogeny of Archosauria (Tetrapoda: Diapsida). *Journal of Systematic Palaeontology* 8(1): 3–47.

Brusatte, S. L., et al. (2010a). The origin and radiation of dinosaurs. *Earth-Science Reviews* 101: 68–100.

Brusca, R. C. and G. J. Brusca (2003). *Invertebrates*, 2nd edition. Sunderland, MA: Sinauer Associates.

Budd, G. (1993). A Cambrian gilled lobopod from Greenland. *Nature* 364: 707–11.

Budd, G. E. (1996). The morphology of *Opabinia regalis* and the reconstruction of the arthropod stem-group. *Lethaia* 29: 1–14.

Budd, G. E. (1998). Arthropod body-plan evolution in the Cambrian with an example from anomalocaridid muscle. *Lethaia* 31: 197–210.

Budd, G. E. (1999). The morphology and phylogenetic significance of *Kerygmachela kierkegaardi* Budd. *Transactions of the Royal Society of Edinburgh: Earth Sciences* 89: 249–90.

Budd, G. E. (2001). Tardigrades as 'stem-group arthropods': the evidence from the Cambrian fauna. *Zoologischer Anzeiger* 240: 265–79.

Budd, G. E. and A. Daley (2011). The lobes and lobopods of *Opabinia regalis* from the middle Cambrian Burgess Shale. *Lethaia* 45(1): 83–95.

Budd, G. E. and S. Jensen (2000). A critical appraisal of the fossil record of the bilaterian phyla. *Biological Reviews* 75: 253–95.

Budd, G. E. and M. J. Telford (2009). The origin and evolution of arthropods. *Nature* 457: 812–17.

Butterfield, N. J. (1990). A reassessment of the enigmatic Burgess Shale fossil *Wiwaxia corrugata* (Matthew) and its relationship to the polychaete *Canadia spinosa* Walcott. *Paleobiology* 16(3): 287–303.

Butterfield, N. J. (2006). Hooking some stem-group 'worms': fossil lophotrochozoans in the Burgess Shale. *BioEssays* 28: 1161–6.

Butterfield, N. J. (2008). An Early Cambrian radula. *Journal of Paleontology* 82(3): 543–54.

Butterfield, N. J. (2009). Modes of pre-Ediacaran multicellularity. *Precambrian Research* 173: 201–11.

Butterfield, N. J. (2009a). Oxygen, animals and oceanic ventilation: an alternative view. *Geobiology* 7: 1–7.

Butterfield, N. J. (2011). Animals and the invention of the Phanerozoic Earth system. *Trends in Ecology and Evolution* 26(2): 81–7.

Callebaut, W. and D. Rasskin-Gutman (2005). *Modularity: Understanding the Development and Evolution of Natural Complex Systems*. Cambridge, MA: MIT Press.

Campbell, K. S. W. and M. W. Bell (1977). A primitive amphibian from the late Devonian of New South Wales. *Alcheringia* 1:369–81.

Canfield, D. E., et al. (2007). Late-Neoptoterozoic deep-ocean oxygenation and the rise of animal life. *Science* 315: 92–5.

Caron, J.-B. (2006). *Banffia constricta*, a putative vetulicolid from the Middle Cambrian Burgess Shale. *Transactions of the Royal Society of Edinburgh, Earth Sciences* 96: 95–111.

Caron, J.-B., et al. (2006). A soft-bodied mollusc with radula from the Middle Cambrian Burgess Shale. *Nature* 442: 159–63.

Caron, J.-B., et al. (2007). Reply to Butterfield on stem-group 'worms': fossil lophotrochozoans in the Burgess Shale. *BioEssays* 29: 200–02.

Caron, J.-B., et al. (2010). Tentaculate fossils from the Cambrian of Canada (British Columbia) and China (Yunnan) interpreted as primitive deuterostomes. *PLoS One* **5**(3): 1–12.

Carrano, M. T. (2006). Body-size evolution in dinosaurs. In M. T. Carrano, T. J. Gaudin, R. W. Blob, and J. R. Wible (eds), *Amniote Paleobiology: Perspectives on the Evolution of Mammals, Birds, and Reptiles*. Chicago, IL: Chicago University Press, 225–68.

Carrier, D. R. (1987). The evolution of locomotor stamina in tetrapods: circumventing a mechanical constraint. *Paleobiology* **13**: 326–41.

Carroll, R. L. (2009). *The Rise of the Amphibians*. Baltimore, MD: Johns Hopkins University Press.

Carroll, R.L. (1988). *Vertebrate Paleontology and Evolution*. New York, NY: Freeman.

Carroll, S. B., et al. (2001). *From DNA to Diversity: Molecular Genetics and the Evolution of Animal Design*. Oxford: Blackwell Science.

Cebra-Thomas, J., et al. (2005). How the turtle got its shell: a paracrine hypothesis of carapace formation. *Journal of Experimental Zoology (Molecular and Developmental Evolution)* **304B**: 558–69.

Chatterjee, S. and J. Templin (2007). Biplane wing planform and flight performance of the feathered dinosaur *Microraptor gui*. *Proceedings of the National Academy of Sciences* **104**(5): 1576–80.

Chen, J.-Y., et al. (1995). A possible Early Cambrian chordate. *Nature* **377**: 720–2.

Chen, J.-Y., et al. (1999). An early Cambrian craniate-like chordate. *Nature* **402**: 518–22.

Chen, J.-Y. (2004). *Dawn of Animal Life*. Nanjing: Jiangsu Science and Technical Press.

Chen, J.-Y., et al. (2004a). A new 'great appendage' arthropod from the Lower Cambrian of China and homology of chelicerate chelicerae and raptorial antero-ventral appendages. *Lethaia* **37**: 3–20.

Chen, J.-Y., et al. (2004b). Small bilaterian fossils from 40–55 million years before the Cambrian. *Science* **305**: 218–22.

Chernoff, B. and P. M. Magwene (1999). Afterword. Morphological integration: forty years later. In E.C. Olson and R. L. Miller (eds), *Morphological Integration*. Chicago, IL: Chicago University Press, 319–48.

Cheverud, J. M., et al. (1997). Pleiotropic effects of individual gene loci on mandibular morphology. *Evolution and Development* **51**: 2006–16.

Cheverud, J. M., et al. (2004). Pleiotropic effects on mandibular morphology II: differential epistasis and genetic variation in morphological integration. *Journal of Experimental Zoology* **302B**: 424–35.

Chiappe, L. M. and G. J. Dyke (2007). The beginnings of birds: recent discoveries, ongoing arguments, and new discoveries. In J. S. Anderson and H.-D. Sues (eds), *Major Transitions in Vertebrate Evolution*. Bloomington, IN: Indiana University Press, 303–6.

Chinsamy-Turan, A., ed. (2012). *Forerunners of Mammals: Radiation, Histology, Biology*. Bloomington, IN: Indiana University Press.

Chinsamy-Turan, A. (2012a). The microstructure of bones and teeth of nonmammalian therapsids. In A. Chinsamy-Turan (ed.), *Forerunners of Mammals: Radiation, Histology, Biology*. Bloomington, IN: Indiana University Press, 65–88.

Chinsamy, A. and W. J. Hillenius (2004). Physiology of nonavian dinosaurs. In D. B. Weishampel, P. Dodson, and H. Osmólska (eds), *The Dinosauria*. Berkeley, CA: University of California Press, 643–59.

Ciampaglio, C. N. (2004). Measuring changes in articulate brachiopod morphology before and after the Permian mass extinction event: do developmental constraints limit morphological innovation? *Evolution and Development* **6**(4): 260–74.

Ciampaglio, C. N., et al. (2001). Detecting changes in morphospace occupation patterns in the fossil record: characterization and analysis of measures of disparity. *Paleobiology* **27**(4): 695–715.

Clack, J. A. (1988). New material of the early tetrapod *Acanthostega* from the Upper Devonian of East Greenland. *Palaeontology* **31**: 699–724.

Clack, J. A. (1994). Earliest known tetrapod braincase and the evolution of the stapes and fenestra ovalis. *Nature* **369**: 392–4.

Clack, J. A. (1998). The neurocranium of *Acanthostega gunnari* Jarvik and the evolution of the otic region in tetrapods. *Zoological Journal of the Linnean Society* **122**: 61–97.

Clack, J. A. (2005). The emergence of early tetrapods. *Palaeogeography, Palaeoclimatology, Palaeoecology* **232**: 167–89.

Clack, J. A. (2006). Devonian climate change, breathing, and the origin of the tetrapod stem. *Integrative and Comparative Biology* **47**(4): 510–23.

Clack, J. A. (2009). The fin to limb transition: new data, interpretations, and hypotheses from paleontology and developmental biology. *Annual Review of Earth and Planetary Science* **37**: 163–79.

Clack, J. A. (2012). *Gaining Ground*, 2nd edition. Bloomington, IN: Indiana University Press.

Clack, J. A., et al. (2003). A uniquely specialised ear in a very early tetrapod. *Nature* **425**: 65–9.

Clarke, A. and H.-O. Pörtner (2010). Temperature, metabolic power and the evolution of endothermy. *Biological Reviews* **85**: 703–27.

Clarke, J. A., et al. (2006). Insight into the evolution of avian flight from a new clade of Early Cretaceous ornithurines from China and the morphology of *Yixianornis grabaui*. *Journal of Anatomy* **208**: 287–308.

Clarkson, E. N. K. (1979). *Invertebrate Palaeontology and Evolution*, 3rd edition. London: Chapman and Hall.

Cleland, C. E. (2002). Methodological and epistemic differences between historical science and experimental science. *Philosophy of Science* **69**: 474–96.

Cleland, T. P., et al. (2011). Histological, chemical, and morphological re-examination of the 'heart' of a small Late Cretaceous *Thescelosaurus*. *Naturwissenschaften* **98**: 203–11.

Coates, M. I. (1996). The Devonian tetrapod *Acanthostega gunnari* Jarvik: postcranial anatomy, basal tetrapod interrelationships and patterns of skeletal evolution. *Transactions of the Royal Society of Edinburgh: Earth Sciences* **87**: 363–421.

Cobbett, A., et al. (2007). Fossils impact as hard as living taxa in parsimony analyses of morphology. *Systematic Biology* **56**: 753–66.

Codd, J. R., et al. (2008). Avian-like breathing mechanics in maniraptoran dinosaurs. *Proceedings of the Royal Society B* **275**: 157–61.

Collins, D. (1996). The 'evolution' of *Anomalocaris* and its classification in the arthropod class Dinocarida (nov) and order Radiodonta (nov). *Journal of Paleontology* **70**: 280–93.

Cong, P.-Y., et al. (2014). New data on the palaeobiology of the enigmatic yunnanozoans from the Chengjiang biota, Lower Cambrian, China. *Palaeontology* **58**: 45–70.

Conway Morris, S. (1977). Aspects of the Burgess Shale fauna with particular reference to the non-arthropod component. *Journal of Paleontology* **51**(2), Supplement Part 3: 7–8.

Conway Morris, S. (1985). The Middle Cambrian metazoan *Wiwaxia corrugata* (Matthew) from the Burgess Shale and *Ogygopsis* Shale, British Columbia. *Philosophical Transactions of the Royal Society B* **307**: 507–82.

Conway Morris, S. (1998). *The Crucible of Creation*. Oxford: Oxford University Press.

Conway Morris, S. and B. Caron (2014). A primitive fish from the Cambrian of North America. *Nature* **512**: 419–22.

Conway Morris, S. and J.-B. Caron (2007). Halwaxiids and the early evolution of the lophotrochozoans. *Science* **315**: 1255–8.

Conway Morris, S. and J.-B. Caron (2012). *Pikaia gracilens* Walcott, a stem-group chordate from the Middle Cambrian of British Columbia. *Biological Reviews* **87**: 480–512.

Conway Morris, S. and J. S. Peel (1990). Articulated halkieriids from the Lower Cambrian of north Greenland. *Nature* **345**: 802–5.

Conway Morris, S. and J. S. Peel (1995). Articulated halkieriids from the Lower Cambrian of north Greenland and their role in early protostome evolution. *Philosophical Transactions of the Royal Society B* **347**: 305–58.

Corfe, I. J. and R. J. Butler (2006). Comment on 'A well-preserved *Archaeopteryx* specimen with theropod features'. *Science* **313**: 1238B.

Cox, C. B. (1969). The problematic Permian reptile *Eunotosaurus*. *Bulletin of the British Museum (Natural History)* **18**: 165–96.

Currie, P. J. and P. J. Chen. (2001). Anatomy of *Sinosauropteryx* prima from Liaoning, northeastern China. *Canadian Journal of Earth Sciences* 38: 1705–22.

Daeschler, E. B., et al. (2006). A Devonian tetrapod-like fish and the evolution of the tetrapod body plan. *Nature* **440**: 757–63.

Daley, A. C., et al. (2009). The Burgess Shale Anomalocaridid *Hurdia* and its significance for early arthropod evolution. *Science* **323**: 1597–600.

Dassow, G. v. and E. Munro (1999). Modularity in animal development and evolution: elements of a conceptual framework for EvoDevo. *Journal of Experimental Zoology (Molecular and Developmental Evolution)* **285**: 307–25.

David, B., et al. (2000). Are homalozoans echinoderms? An answer from the extraxial–axial theory. *Paleobiology* **26**: 529–55.

Davidson, E. H. and D. H. Erwin (2006). Gene regulatory networks and the evolution of animal body plans. *Science* **311**: 796–800.

Davis, M. C. (2013). The deep homology of the autopod: insights from hox gene regulation. *Integrative and Comparative Biology* **53**(2): 224–32.

de Beer, G. R. (1958). *Embryos and Ancestors*. Oxford: Oxford University Press.

deBragga, M. and O. Rieppel (1997). Reptile phylogeny and the interrelationships of turtles. *Zoological Journal of the Linnean Society* **120**: 281–354.

Dececchi, T. A. and H. C. E. Larsson (2011). Assessing arboreal adaptations of bird antecedents: testing the ecological setting of the origin of the avian flight stroke. *PLoS One* **6**(8): e22292.

Delsuc, F., et al. (2006). Tunicates and not cephalochordates are the closest living relatives of vertebrates. *Nature* **439**: 965–8.

Dial, K. P. (2003). Wing-assisted incline running and the evolution of flight. *Science* **299**: 402–4.

Dial, K. P., et al. (2006). What use is half a wing in the ecology and evolution of birds? *Bioscience* **56**(5): 437–45.

Dohrmann, M. and G. Wörheide (2013). Novel scenarios of early animal evolution—is it time to rewrite the textbooks? *Integrative and Comparative Biology* **53**(3): 503–11.

Donaghue, M. J., et al. (1989). The importance of fossils in phylogeny reconstruction. *Annual Review of Ecology and Systematics* **20**: 431–60.

Donoghue, P. C. J. and M. J. Benton (2007). Rocks and clocks: calibrating the tree of life using fossils and molecules. *Trends in Ecology and Evolution* **22**(8): 424–31.

Donoghue, P. C. J. and J. N. Keating (2014). Early vertebrate evolution. *Palaeontology* **57**(5): 879–93.

Donoghue, P. C. J. and M. A. Purnell (2009). Distinguishing heat from light in debate over controversial fossils. *BioEssays* **31**: 178–89.

Downs, J. P., et al. (2008). The cranial endoskeleton of *Tiktaalik roseae*. *Nature* **455**: 925–9.

Dullemeijer, P. (1980). Functional morphology and evolutionary biology. *Acta Biotheoretica* **29**: 151–250.

Dunn, C. W., et al. (2008). Broad phylogenetic sampling improves resolution of the animal tree of life. *Nature* **452**: 745–50.

Dzik, J. (2010). Brachiopod identity of the alleged monoplacophoran ancestors of cephalopods. *Malacologia* **52**: 97–113.

Edgecombe, G. D. (2010). Arthropod phylogeny: an overview from the perspectives of morphology, molecular data and the fossil record. *Arthropod Structure and Development* **39**: 74–87.

Edgecombe, G. D. (2010a). Palaeomorphology: fossils and the inference of cladistic relationships. *Acta Zoologica* **91**: 72–80.

Edgecombe, G. D., et al. (2011). Higher-level metazoan relationships: recent progress and remaining questions. *Organisms, Diversity and Evolution* **11**: 151–72.

Edgecombe, G. D. and D. A. Legg (2014). Origins and early evolution of arthropods. *Palaeontology* **57**: 457–68.

Eibye-Jacobsen, D. (2004). A reevaluation of *Wiwaxia* and the polychaetes of the Burgess Shale. *Lethaia* **37**: 317–35.

Elicki, O. and S. Gursu (2009). First record of *Pojetaia runnegari* Jell, 1980 and *Fordilla* Barrande, 1881 from the Middle East (Taurus Mountains, Turkey) and critical review of Cambrian bivalves. *Palaeontologische Zeitschrift* **83**(2): 267–91.

Else, P. L. (2013). Dinosaur lactation? *Journal of Experimental Biology* **216**: 347–51.

Erickson, G., et al. (2009). Was dinosaurian physiology inherited by birds? Reconciling slow growth in *Archaeopteryx*. *PLoS One* **4**(10 e7390): 1–9.

Erwin, D. H. (2007). Disparity: morphological pattern and developmental context. *Palaeontology* **50**(1): 57–73.

Erwin, D. H. and E. H. Davidson (2009). The evolution of hierarchical gene regulatory networks. *Nature Reviews Genetics* **10**: 141–8.

Erwin, D. H. and J. W. Valentine (2013). *The Cambrian Explosion: The Construction of Animal Biodiversity*. Greenwood Village, CO: Roberts and Company.

Erwin, D. H., et al. (1987). A comparative study of diversification events: the early Paleozoic versus the Mesozoic. *Evolution* **41**(6): 1177–86.

Erwin, D. H., et al. (2011). The Cambrian conundrum: early divergence and later ecological success in the early history of animals. *Science* **334**: 1091–7.

Fang, Z.-J. (2006). An introduction to Ordovician bivalves of southern China, with a discussion of the early evolution of the Bivalvia. *Geological Journal* **41**: 303–28.

Fedonkin, M. and B. M. Waggoner (1997). The Late Precambrian fossil *Kimberella* is a mollusc-like bilaterian organism. *Nature* **388**: 868–71.

Fedonkin, M. A., et al. (2007). New data on *Kimberella*, the Vendian mollusc-like organism (White Sea region, Russia): palaeoecological and evolutionary implications. In P. Vickers-Rich and P. Komarower (eds), *The Rise and Fall of the Ediacaran Biota. Geological Society Special Publication No. 286*. London: The Geological Society, 157–79.

Fisher, P. E., et al. (2000). Cardiovascular evidence for an intermediate or higher metabolic rate in an ornithischian dinosaur. *Science* **288**: 503–5.

Fisher, R. A. (1930). *The Genetic Theory of Natural Selection*. Oxford: Oxford University Press.

Fong, J. J., et al. (2012). A phylogenomic approach to vertebrate phylogeny supports a turtle-archosaur affinity and a possible paraphyletic Lissamphibia. *PLoS One* **7**(11 e48990): 1–14.

Foote, M. (1997). The evolution of morphological diversity. *Annual Review of Ecology and Systematics* **28**: 129–52.

Foote, M. (2003). Estimating completeness in the fossil record. In D. E. G. Briggs and P. R. Crowther (eds), *Palaeobiology II*. Oxford: Blackwell Publishing, 500–4.

Frei, R., et al. (2009). Fluctuations in Precambrian atmospheric oxygenation recorded by chromium isotopes. *Nature* **461**: 250–3.

Fristrup, K. M. (2001). A history of character concepts in evolutionary biology. In G. P. Wagner (ed.), *The Character Concept in Biology*. San Diego, CA: Academic Press, 15–37.

Gaffney, E. S. (1990). The comparative osteology of the Triassic turtle *Proganochelys*. *Bulletin of the American Museum of Natural History* **194**: 1–263.

Galis, F. (2001). Key innovations and radiations. In G. P. Wagner (ed.), *The Character Concept in Evolutionary Biology*. San Diego, CA: Academic Press, 583–607.

Garcia-Bellido, D. C., et al. (2014). A new vetulicolian from Australia and its bearing on the chordate affinities of an enigmatic Cambrian group. *BMC Evolutionary Biology* **14**: 214–26.

Garstang, W. (1894). Preliminary note on a new theory of the phylogeny of the Chordata. *Zoologischer Anzeiger* **17**: 122–5.

Garstang, W. (1928). The morphology of the Tunicata, and its bearing on the phylogeny of the Chordata. *Quarterly Journal of Microscopy* **72**: 51–187.

Gauthier, J. (1986). Saurischian monophyly and the origin of birds. In K. Padian (ed.), *The Origin of Birds and the Evolution of Flight*. Berkeley, CA: Academy of Science, 1–55.

Gauthier, J., et al. (1988). Amniote phylogeny and the importance of fossils. *Cladistics* **4**: 105–209.

Gavrilets, S. (1997). Evolution and speciation on holey adaptive landscapes. *Trends in Ecology and Evolution* **12**: 307–12.

Gee, H. (2000). *In Search of Deep Time*. New York, NY: The Free Press.

Gehling, J. G., et al. (2011). The geological context of the Lower Cambrian (Series 2) Emu Bay Shale Lagerstätte and adjacent stratigraphic units, Kangaroo Island, Australia. *Australian Journal of Earth Sciences* **58**: 243–57.

George, D. and A. Blieck (2011). Rise of the earliest tetrapods: an Early Devonian origin from marine environment. *PLoS One* **6**(7): e22136.

Ghiselin, M. T. (1974). A radical solution to the species problem. *Systematic Zoology* **23**: 536–44.

Gilchrist, G. W. and J. G. Kingsolver (2001). Is optimality over the hill? In S. H. Orzack and E. Sober (eds), *Adaptationism and Optimality*. Cambridge: Cambridge University Press, 219–41.

Gingerich, P. D., et al. (1994). New whale from the Eocene of Pakistan and the origin of whale swimming. *Nature* **368**: 844–7.

Gingerich, P. D. and M. D. Uhen (1996). *Ancalecetus simonsi*, a new dorudontine archaeocete (Mammalia, Cetacea) from the early late Eocene of Wadi Hitan, Egypt. *Contributions from the Museum of Paleontology University of Michigan* **29**: 359.

Gingerich, P. D., et al. (2001). Origin of whales from early artiodactyls: hands and feet of Eocene Protocetidae from Pakistan. *Science* **293**: 2239–42.

Gingerich, P. D., et al. (2009). New protocetid whale from the Middle Eocene of Pakistan: birth on land, precocial development, and sexual Dimorphism. *PLoS One* **4**(2, e4366): 1–20.

Glaessner, M. E. (1984). *The Dawn of Animal Life*. Cambridge: Cambridge University Press.

Glaessner, M. E. and M. Wade (1966). The Late Precambrian fossils from Ediacara, South Australia. *Palaeontology* **9**: 599–628.

Goddéris, Y., et al. (2012). Tectonic control of continental weathering, atmospheric CO_2, and climate over Phanerozoic time. *C.R. Geoscience* **344**: 652–62.

Goh, K.-I., et al. (2007). The human disease network. *Proceedings of the National Academy of Sciences* **104**(21): 8685–90.

Goodnight, C. J. (2012). Wright's shifting balance theory and factors affecting the probability of peak shifts. In E. I. Svensson and R. Calsbeek (eds), *The Adaptive Landscape in Evolutionary Biology*. Oxford: Oxford University Press, 74–86.

Goodrich, E. S. (1930). *Studies on the Structure and Development of Vertebrates*. London: Macmillan.

Goswami, A. (2006). Cranial modularity shifts during mammalian evolution. *American Naturalist* **168**(2): 270–80.

Goswami, A. and P. D. Polly (2010). The influence of modularity on cranial morphological disparity in carnivora and Primates (Mammalia). *PLoS One* **5**(3, e9517): 9511–18.

Gould, S. J. (1977). *Ontogeny and Phylogeny*. Cambridge, MA: Harvard University Press.

Gould, S. J. (1983). Irrelevance, submission and partnership: the changing role of paleontology in Darwin's three centennials, and a modest proposal for macroevolution. In D. S. Bendall (ed.), *Evolution from Molecules to Man*. Cambridge: Cambridge University Press, 347–66.

Gould, S. J. (1989). *Wonderful Life: The Burgess Shale and the Nature of History*. London: Hutchinson Radius.

Gould, S. J. (1991). The disparity of the Burgess Shale arthropod fauna and the limits of cladistic analysis: why we must strive to quantify morphospace. *Paleobiology* **17**: 411–23.

Gould, S. J. (2000). Of coiled oysters and big brains: how to rescue the terminology of heterochrony, now gone astray. *Evolution and Development* **2**(5): 241–48.

Gregory, W. K. (1910). The orders of mammals. *Bulletin of the American Museum of Natural History* **27**: 1–524.

Grene, M. (1958). Two evolutionary theories. *British Journal of the History of Science* **9**: 110–27 and 185–93.

Gribbin, J. R., et al. (2009). *In Search of Gaia*. Princeton, NJ: Princeton University Press.

Griswold, C. K. (2006). Pleiotropic mutation, modularity and evolvability. *Evolution and Development* **8**(1): 81–93.

Haeckel, E. (1866). *Generelle Morphologie der Organismen: Allgemeine Grundzüge der organischen Formen-Wissenschaft, mechanisch begründet durch die von Charles Darwin reformite Descendenz-Theorie*, 2 volumes. Berlin: Georg Reimer.

Haeckel, E. (1874). Die Gastraea-Theorie, die phylogenetische Classification des Thierrreisches und die Homologie der Keimblatter. *Jenaische Zeitschrift fur Naturwissenschaft* **8**: 1–55.

Halanych, K. M., et al. (1995). Evidence from 18S ribosomal DNA that the lophophorates are protostome animals. *Science* **267**: 1641–3.

Haldane, J. B. S. (1957). The cost of natural selection. *Journal of Genetics* **55**: 511–24.

Hall, B. K. (1995). Homology and development. *Evolutionary Biology* **28**: 1–37.

Hallgrímsson, B. and B. K. Hall (eds), (2011). *Epigenetics: Linking Genotype and Phenotype in Development and Evolution*. Berkeley, CA: California University Press.

Hanken, J. and D. B. Wake (1993). Miniaturization of body size: organismal consequences and evolutionary significance. *Annual Review of Ecology and Systematics* **24**: 501–19.

Hansen, T. F. (2003). Is modularity necessary for evolvability? Remarks on the relationship between pleiotropy and evolution. *Biosystems* **69**: 83–94.

Hansen, T. F. (2012). Adaptive landscapes and macroevolutionary dynamics. In E. I. Svensson and R. Calsbeek (eds), *The Adaptive Landscape in Evolutionary Biology*. Oxford: Oxford University Press, 205–26.

Haszprunar, G. (2000). Is the Aplacophora monophyletic? A cladistic point of view. *American Malacological Bulletin* **15**(2): 115–30.

Hausdorf, B. (2011). Progress towards a general species concept. *Evolution* **65**(4): 923–31.

Hayes, J. M. and J. R. Waldbauer (2006). The carbon cycle and associated redox processes through time. *Philosophical Transactions of the Royal Society B* **361**: 931–50.

He, J. and M. W. Deem (2010). Hierarchical evolution of animal body plans. *Developmental Biology* **337**: 157–61.

Hedges, S. B. and S. Kumar (eds), (2009). *The Timetree of Life*. Oxford: Oxford University Press.

Helmkampf, M., et al. (2008). Phylogenomic analyses of lophophorates (brachiopods, phoronids and bryozoans) confirm the Lophotrochozoa concept. *Proceedings of the Royal Society B* **275**: 1927–33.

Hillenius, W. J. (1994). Turbinates in therapsids: evidence for late Permian origin of mammalian endothermy. *Evolution* **48**: 207–29.

Hoffman, P. F., et al. (1998). A Neoproterozoic snowball Earth. *Science* **281**: 1342–6.

Holman, E. W. (1996). The independent variable in the early origin of higher taxa. *Journal of Theoretical Biology* **181**: 85–94.

Hooper, J. E. and M. P. Scott (2005). Communicating with Hedgehogs. *Nature Reviews of Molecular and Cellular Biology* **6**: 306–17.

Hopson, J. A. (1979). Paleoneurology. In C. Gans, G. G. Northcutt, and P. Ulinski (eds), *Biology of the Reptilia*, volume 9. London: Academic Press, 39–146.

Hopson, J. A. (2012). The role of foraging mode in the origin of therapsids: implications for the origin of mammalian endothermy. *Fieldiana: Life and Earth Sciences* **5**: 126–48.

Hou, X. G., et al. (2007). *The Cambrian Fossils of Chengjiang, China*. Oxford: Blackwell Publishing.

Hu, D., et al. (2009). A pre-*Archaeopteryx* troodontid theropod from China with long feathers on the metatarsus. *Nature* **461**: 640–3.

Hunter, J. P. (1998). Key innovations and the ecology of macroevolution. *Trends in Ecology and Evolution* **13**: 31–6.

Hunter, J. P. and J. Jernvall (1995). The hypocone as a key innovation in mammalian evolution. *Proceedings of the National Academy of Sciences* **92**(23): 10718–22.

Hutchinson, J. R. and V. Allen (2009). The evolutionary continuum of limb function from early theropods to birds. *Naturwissenschaften* **96**: 423–48.

Huttenlocker, A. K. and E. Rega (2012). The palaeobiology and bone microstructure of pelycosaurian-grade synapsids. In A. Chinsamy-Turan (ed.), *Forerunners of Mammals: Radiation, Histology, Biology*. Bloomington, IN: Indiana University Press, 91–119.

Huxley, T. H. (1868). Remarks upon *Archaeopteryx lithographica*. *Proceedings of the Royal Society* **16**: 243–8.

Huxley, T. H. (1886). On the animals which are most intermediate between birds and reptiles. *Annals and Magazine of Natural History Series 4* **2**: 66–75.

Hyman, L. H. (1951). *The Invertebrates. Volume 2, Platyhelminthes and Rhynchocoela*. New York, NY: McGraw-Hill.

Ingham, P. W., et al. (2011). Mechanisms and functions of Hedgehog signalling across the Metazoa. *Nature Reviews Genetics* **12**: 393–406.

Irmis, R. B. (2011). Evaluating hypotheses for the early diversification of dinosaurs. *Earth and Environmental Science Transactions of the Royal Society of Edinburgh* **101**: 397–426.

Ivakhnenko, M. F. (1999). Biarmosuches from the Ocher faunal assemblage of eastern Europe. *Paleontological Journal* **33**: 289–96.

Ivantsov, A. Y. (2009). New reconstruction of *Kimberella*, problematic Vendian metazoan. *Palaeontological Journal* **43**(6): 601–11.

Jablonski, D. (2007). Scale and hierarchy in macroevolution. *Palaeontology* **50**(1): 87–109.

Jablonski, D. and D. J. Bottjer (1990). The origin and diversification of major groups: environmental patterns and macroevolutionary lags. In P. D. Taylor and G. P. Larwood (eds), *The Systematics Association Special Volume No. 42. Major Evolutionary Radiations*. Oxford: Clarendon Press, 17–57.

Jamniczky, H. A. and B. Hallgrímsson (2011). Modularity of the skull and cranial vasculature of laboratory mice: implications for the evolution of complex phenotypes. *Evolution and Development* **13**(1): 28–37.

Jarvik, E. (1980). *Basic Structure and Evolution of Vertebrates, Volumes 1 and 2*. New York, NY: Academic Press.

Jarvik, E. (1996). The Devonian tetrapod *Ichthyostega*. *Fossils and Strata* **40**: 1–206.

Jasinoski, S. C., et al. (2010). Functional implications of dicynodont cranial suture morphology. *Journal of Morphology* **271**: 705–28.

Jeffares, B. (2010). Guessing the future of the past. *Biology and Philosophy* **25**: 125–42.

Jefferies, R. P. S. (1979). The origin of chordates—a methodological essay. In M. R. House (ed.), *The Origin of the Major Invertebrate Groups*. London: Academic Press, 443–77.

Jefferies, R. P. S. (1986). *The Ancestry of Vertebrates*. London: British Museum.

Jefferies, R. P. S., et al. (1996). The early phylogeny of chordates and echinoderms and the origin of chordate left-right asymmetry and bilateral symmetry. *Acta Zoologica* **77**: 101–22.

Jenkins, F. A. Jr and F. R. Parrington (1976). The postcranial skeleton of the Triassic mammals *Eozostrodon*, *Megazostrodon* and *Erythrotherium*. *Philosophical Transactions of the Royal Society B* 273: 387–431.

Ji, Q. and S.-A. Ji (1996). On the discovery of the earliest fossil bird in China (*Sinosauropteryx* gen. nov.) and the origin of birds. *Chinese Geology* 233: 30–3.

Jones, D. S. and S. J. Gould (1999). Direct measurement of age in fossil *Gryphaea*: the solution to a classic problem in heterochrony. *Paleobiology* 25: 158–87.

Kauffman, S. A. (1993). *The Origins of Order*. New York, NY: Oxford University Press.

Kemp, T. S. (1978). Stance and gait in the hindlimb of a therocephalian mammal-like reptile. *Journal of Zoology* 186: 143–61.

Kemp, T. S. (1979). The primitive cynodont *Procynosuchus*: functional anatomy of the skull and relationships. *Philosophical Transactions of the Royal Society B* 285: 73–122.

Kemp, T. S. (1980). Aspects of the structure and functional anatomy of the Middle Triassic cynodont *Luangwa*. *Journal of Zoology* 191: 193–239.

Kemp, T. S. (1980a). The primitive cynodont *Procynosuchus*: structure, function and evolution of the postcranial skeleton. *Philosophical Transactions of the Royal Society B* 288: 217–58.

Kemp, T. S. (1982). *Mammal-like Reptiles and the Origin of Mammals*. London: Academic Press.

Kemp, T. S. (1985). Models of diversity and phylogenetic reconstruction. In R. Dawkins and M. Ridley (eds), *Oxford Surveys in Evolutionary Biology*, Volume 2. Oxford: Oxford University Press, 135–58.

Kemp, T. S. (1985a). Synapsid reptiles and the origin of higher taxa. *Special Papers in Palaeontology* 33: 175–84.

Kemp, T. S. (1986). The skeleton of a baurioid therocephalian therapsid from the Lower Triassic (*Lystrosaurus* zone) of South Africa. *Journal of Vertebrate Paleontology* 6: 215–32.

Kemp, T. S. (1988). The origin of mammals: observed pattern and inferred process. *L'evolution dans sa realité et ses diverses modalites*. Paris: Fondation Singer-Polignac and Masson, 65–91.

Kemp, T. S. (1999). *Fossils and Evolution*. Oxford: Oxford University Press.

Kemp, T. S. (2005). *The Origin and Evolution of Mammals*. Oxford: Oxford University Press.

Kemp, T. S. (2006). The origin and early radiation of the therapsid mammal-like reptiles: a paleobiological hypothesis. *Journal of Evolutionary Biology* 19: 1231–47.

Kemp, T. S. (2006a). The origin of mammalian endothermy: a paradigm for the evolution of complex biological structure. *Zoological Journal of the Linnean Society* 147: 473–88.

Kemp, T. S. (2007). Acoustic transformer function of the postdentary bones and quadrate of a nonmammalian cynodont. *Journal of Vertebrate Paleontology* 27(2): 431–41.

Kemp, T. S. (2007a). The origin of higher taxa: macroevolutionary processes, and the case of the mammals. *Acta Zoologica* 88: 3–22.

Kemp, T. S. (2007b). The concept of correlated progression as the basis of a model for the evolutionary origin of major new taxa. *Proceedings of the Royal Society B* 274: 1667–73.

Kemp, T. S. (2009). The endocranial cavity of a nonmammalian eucynodont *Chiniquodon theotenicus* and its implications for the origin of the mammalian brain. *Journal of Vertebrate Paleontology* 29(4): 1188–98.

Kemp, T. S. (2009a). Phylogenetic interrelationships and pattern of evolution of the therapsids: testing for polytomy. *Palaeontologia Africana* 44: 1–12.

Kemp, T. S. (2010). New perspectives on the evolution of Late Palaeozoic and Mesozoic terrestrial tetrapods. In S. Badyopadhyay (ed.), *New Aspects of Mesozoic Diversity*. Heidelberg: Springer, 1–26.

Kemp, T. S. (2012). The origin and radiation of the therapsids. In A. Chinsamy-Turan (ed.), *Forerunners of Mammals: Radiation, Histology, Biology*. Bloomington, IN: Indiana University Press, 1–28.

Kielan-Jaworowska, Z., et al. (2004). *Mammals from the Age of Dinosaurs: Origins, Evolution, and Structure*. New York, NY: Columbia University Press.

Kimura, M. (1983). *The Neutral Theory of Molecular Evolution*. Cambridge: Cambridge University Press.

King, J. L. and T. H. Jukes (1969). Non-Darwinian evolution. *Science* 164: 788–98.

Kirkland, J. I., et al. (2005). A primitive therizinosauroid dinosaur from the Early Cretaceous of Utah. *Nature* 435: 84–7.

Klingenberg, C. P. (1998). Heterochrony and allometry: the analysis of evolutionary change in ontogeny. *Biological Reviews* 73: 79–123.

Klingenberg, C. P. (2008). Morphological integration and developmental modularity. *Annual Review of Ecology, Evolution, and Systematics* 39: 115–32.

Kluge, B., et al. (2005). Anatomical and molecular reinvestigation of lamprey endostyle development provides new insight into thyroid gland evolution. *Development, Genes, and Evolution* 215: 32–40.

Kocot, K. M., et al. (2011). Phylogenomics reveals deep molluscan relationships. *Nature* 477: 452–57.

Kolata, D. R., et al. (1991). The youngest carpoid: occurrence, affinities and life mode of a Pennsylvanian (Morrowan) mitrate from Oklahoma. *Journal of Paleontology* 65: 844–55.

Kröger, B. (2007). Some lesser known features of the ancient cephalopod order Ellesmerocerida (Nautiloidea, Cephalopoda). *Palaeontology* 50(3): 565–72.

Kröger, B., et al. (2011). Cephalopod origin and evolution: a congruent picture emerging from fossils, development and molecules. *BioEssays* **33**: 602–13.

Kühl, G., et al. (2009). A great appendage arthropod with a radial mouth from the Lower Devonian Hunsrück Slate, Germany. *Science* **323**: 771–3.

Kumar, S. and S. B. Hedges (eds), (2009). *The Timetree of Life*. Oxford: Oxford University Press.

Kundrat, M. (2007). Avian-like attributes of a virtual brain model of the oviraptorid theropod *Conchoraptor gracilis*. *Naturwissenschaften* **94**: 499–504.

Laass, M., et al. (2010). New insights into the respiration and metabolic physiology of *Lystrosaurus*. *Acta Zoologica* **92**: 363–71.

Lacalli, T. C. (2002). Vetulicolians—are they deuterostomes? chordates? *BioEssays* **24**: 208–11.

Laflamme, M., et al. (2012). The end of the Ediacara biota: extinction, biotic replacement, or Cheshire Cat? *Gondwana Research* **23**: 558–73.

Langer, M. C., et al. (2010). The origin and early evolution of dinosaurs. *Biological Reviews* **85**: 55–110.

Laurin, M. and P. D. Cantino (2004). First international phylogenetic nomenclature meeting. *Zoologica Scripta* **33**(5): 475–9.

Laurin, M. and P. D. Cantino (2006). Second Meeting of the International Society for Phylogenetic Nomenclature: A Report. *Zoologica Scripta* **36**(1): 109–17.

Laurin, M. and R. R. Reisz (1995). A reevaluation of early amniote phylogeny. *Biological Journal of the Linnean Society* **113**: 165–223.

Lebedev, O. A. and J. A. Clack (1993). Upper Devonian tetrapods from Andreyevka, Tula Region, Russia. *Palaeontology* **36**(3): 721–34.

Lebedev, O. A. and M. I. Coates (1995). The postcranial skeleton of the Devonian tetrapod *Tulerpeton curtum* Lebedev. *Zoological Journal of the Linnean Society* **114**: 307–48.

Lee, M. S. Y. (1996). Correlated progression and the origin of turtles. *Nature* **379**: 812–15.

Lee, M. S. Y. (1997). Pareiasaur phylogeny and the origin of turtles. *Zoological Journal of the Linnean Society* **120**: 197–280.

Lee, M. S. Y. and T. H. Worthy (2011). Likelihood reinstates *Archaeopteryx* as a primitive bird. *Biology Letters* **8**: 299–303.

Lee, M. S. Y., et al. (2014). Sustained miniaturization and anatomical innovation in the dinosaurian ancestors of birds. *Science* **345**: 562–66.

Lefevre, U., et al. (2014). A new long-tailed basal bird from the Lower Cretaceous of north-eastern China. *Biological Journal of the Linnean Society* 113(3) Special issue S1: 790–804.

Legg, D. (2013). Multi-segmented arthropods from the Middle Cambrian of British Columbia (Canada). *Journal of Paleontology* **87**: 493–501.

Legg, D. A., et al. (2012). Cambrian bivalved arthropod reveals origin of arthrodization. *Proceedings of the Royal Society B* **279**: 4699–704.

Legg, D. A., et al. (2013). Arthropod fossil data increase congruence of morphological and fossil phylogenies. *Nature Communications* **4**, **2485**: 1–7.

Lenski, R. E., et al. (2003). The evolutionary origin of complex features. *Nature* **423**: 139–44.

Li, C., et al. (2008). An ancestral turtle from the Late Triassic of southwestern China. *Nature* **456**: 497–501.

Liang, D., et al. (2013). One thousand two hundred and ninety nuclear genes from a genome-wide survey support lungfishes as the sister group of tetrapods. *Molecular Biology and Evolution* **30**(8): 1803–7.

Liu, J. and J. A. Dunlop (2014). Cambrian lobopodians: a review of recent progress in our understanding of their morphology and evolution. *Palaeogeography, Palaeoclimatology, Palaeoecology* **2014**: 398.

Liu, J., et al. (2009). New basal synapsid supports Laurasian origin for therapsids. *Acta Palaeontologica Polonica* **54**(3): 393–400.

Liu, J., et al. (2011). An armoured Cambrian lobopodian from China with arthropod-like appendages. *Nature* **470**: 526–30.

Long, J. and P. Schouten (2008). *Feathered Dinosaurs*. New York, NY: Oxford University Press.

Long, J. A. and K. J. McNamara (1995). Heterochrony in dinosaur evolution. In K. J. McNamara (ed.), *Evolutionary Change and Heterochrony*. New York, NY: Wiley, 151–68.

Longrich, N. (2006). Structure and function of hindlimb features in *Archaeopteryx lithographica*. *Paleobiology* **32**(3): 417–31.

Lorenz, D. M., et al. (2011). The emergence of modularity in biological systems. *Physics of Life Reviews* **8**: 129–60.

Lovegrove, B. (2011). The evolution of endothermy in Cenozoic mammals: a plesiomorphic–apomorphic continuum. *Biological Reviews* **87**: 128–62.

Lovelock, J. (1979). *Gaia: A New Look at Life on Earth*, 3rd edition. Oxford: Oxford University Press.

Lull, R. S. (1918). The pulse of life. In J. Barrell (ed.), *The Evolution of the Earth*. New Haven, CT: Yale University Press, 109–46.

Luo, Z.-X., et al. (2011). A Jurassic eutherian mammal and divergence of marsupials and placentals. *Nature* **476**: 442–5.

Lyson, T. R. and W. G. Joyce (2012). Evolution of the turtle bauplan: the topological relationship of the scapula relative to the ribcage. *Biology Letters* **6**(6): 830–3.

Lyson, T. R., et al. (2011). Transitional fossils and the origin of turtles. *Biology Letters* **6**: 830–3.

Madar, S. I., et al. (2002). Additional holotype remains of *Ambulocetus natans* (Cetacea, Ambulocetidae) and their implications for locomotion in early whales. *Journal of Vertebrate Paleontology* **22**(2): 405–22.

Mallat, J. and N. Holland (2013). *Pikaia gracilens* Walcott: stem chordate, or already specialised in the Cambrian. *Journal of Experimental Zoology (Molecular and Developmental Evolution)* **320B**: 247–71.

Maloof, C. M., et al. (2010). The earliest Cambrian record of animals and ocean geochemical change. *Geological Society of America Bulletin* **122**: 1731–74.

Manton, S. and D. T. Anderson (1979). Polyphyly and the evolution of arthopods. In M. R. House (ed.), *The Origin of Major Invertebrate Groups*. London: Academic Press, 269–321.

Markey, M. J. and C. R. Marshall (2007). Terrestrial-style feeding in a very early aquatic tetrapod is supported by evidence from experimental analysis of suture morphology. *Proceedings of the National Academy of Sciences* **104**(17): 7134–8.

Marks, C. O. (2007). The causes of variation in tree seedling traits: the roles of environmental selection versus chance. *Evolution* **61**: 455–69.

Marks, C. O. and M. J. Lechowicz (2006a). A holistic tree seedling model for the investigation of functional trait diversity. *Ecological Modelling* **193**: 141–81.

Marks, C. O. and M. J. Lechowicz (2006b). Alternative designs and the evolution of functional diversity. *The American Naturalist* **167**(1): 55–66.

Marroig, G., et al. (2009a). The evolution of modularity in the mammalian skull: I: morphological integration patterns and magnitudes. *Evolutionary Biology* **36**: 118–35.

Marroig, G., et al. (2009b). The evolution of modularity in the mammalian skull: II: evolutionary consequences. *Evolutionary Biology* **36**: 136–48.

Marshall, C. R. and J. W. Valentine (2010). The importance of preadapted genomes in the origin of the animal body plans and the Cambrian explosion. *Evolution* **64**(5): 1189–201.

Martin, K. L. and A. L. Carter (2013). Brave new propagules: terrestrial embryos in anamniotic eggs. *Integrative and Comparative Biology* **53**(2): 233–47.

Matus, D. Q., et al. (2008). The Hedgehog family of the cnidarian, *Nematostella vectensis*, and implications for understanding metazoan Hedgehog pathway evolution. *Developmental Biology* **313**: 501–18.

Mayhew, P. J., et al. (2008). A long-term association between global temperature and biodiversity, origination and extinction in the fossil record. *Proceedings of the Royal Society B* **275**: 47–53.

Mayhew, P. J., et al. (2012). Biodiversity tracks temperature over time. *Proceedings of the National Academy of Sciences* **109**(38): 15141–5.

Mayr, E. (1969). *Principles of Systematic Zoology*. Cambridge, MA: Harvard University Press.

Mayr, G., et al. (2005). A well-preserved *Archaeopteryx* specimen with theropod features. *Science* **310**: 1483–86.

Mazurek, D. and M. Zatoń (2011). Is *Nectocaris pteryx* a cephalopod? *Lethaia* **44**: 2–4.

McKenna, M. C. and S. K. Bell (1997). *Classification of Mammals above the Species Level*. New York, NY: Columbia University Press.

McNamara, K. J. and M. L. McKinney (2005). Heterochrony, disparity, and macroevolution. *Paleobiology* **31**(2): 17–26.

Meyer, A. and R. Zardoya (2003). Recent advances in the (molecular) phylogeny of vertebrates. *Annual Review of Ecology, Systematics and Evolution* **34**: 311–38.

Minelli, A. (1993). *Biological Systematics: The State of the Art*. London: Chapman and Hall.

Minelli, A. (2009). *Perspectives in Animal Phylogeny and Evolution*. Oxford: Oxford University Press.

Monteiro, L. R. and M. R. Nogueira (2009). Adaptive radiations, ecological specialization, and the evolutionary integration of complex morphological structures. *Evolution* **64**(3): 724–44.

Monteiro, L. R., et al. (2005). Evolutionary integration and morphological diversification in complex morphological structures: mandible shape divergence in spiny rates (Rodentia, Echimyidae). *Evolution and Development* **7**(5): 429–39.

Moustakas, J. E. (2008). Development of the carapacial ridge: implications for the evolution of genetic networks in turtle shell development. *Evolution and Development* **10**(1): 29–36.

Murdock, D. J. E. and P. C. J. Donaghue (2011). Evolutionary origins of animal skeletal biomineralization. *Cells Tissues Organs* **194**: 98–102.

Nagashima, H., et al. (2009). Evolution of the turtle body plan by the folding and creation of new muscle connections. *Science* **325**: 193–54.

Nardin, E., et al. (2013). Modelling the early Paleozoic long-term climatic trend. *GSA Bulletin* **123**: 1181–92.

Narkiewicz, M. and G. J. Retallack (2014). Dolomitic paleosols in the lagoonal tetrapod track-bearing succession of Holy Cross Mountains (Middle Devonian, Poland). *Sedimentary Geology* **299**: 74–87.

Near, T. J. (2009). Conflict and resolution between phylogenies inferred from molecular and phenotypic data sets for hagfishes, lampreys, and gnathostomes. *Journal of Experimental Zoology (Molecular and Developmental Evolution)* **312B**: 749–61.

Nelson, G. and N. Platnick (1984). Systematics and evolution. In M.-W. Ho and P. T. Saunders (eds), *Beyond Neo-Darwinism*. London: Academic Press, 143–58.

Nesbitt, S. J. (2011). The early evolution of archosaurs: relationships and the origin of major clades. *Bulletin of the American Museum of Natural History* **352**: 1–288.

Neufeld, C. J. and A. R. Palmer (2011). Learning, developmental plasticity, and the rate of morphological evolution. In B. Hallgrímson and B. K. Hall (eds), *Epigenetics: Linking Genotype and Phenotype in Development and Evolution*. Berkeley, CA: University of California Press, 337–56.

Newman, S. A. (2010). Dynamical patterning modules. In M. Pigliucci and G. B. Müller (eds), *Evolution—The Extended Synthesis*. Cambridge, Massachusetts: MIT Press, 281–306.

Newman, S. A., et al. (2006). Before programs: the physical origination of multicellular forms. *International Journal of Developmental Biology* **50**: 289–99.

Nielsen, C. (2008). Six major steps in animal evolution— are we derived sponge larvae? *Evolution and Development* **10**: 241–57.

Niedźwiedzki, G., et al. (2010). Devonian trackways from the early Middle Devonian period of Poland. *Nature* **463**: 43–8.

Niklas, K. J. (1995). Morphological evolution through complex domains of fitness. In W. M. Fitch and F. J. Ayala (eds), *Tempo and Mode in Evolution: Genetics and Paleontology 50 Years After Simpson*. Washington, DC: National Academy Press, 145–68.

Niklas, K. J. (1997). *The Evolutionary Biology of Plants*. Chicago, IL: Chicago University Press.

Niklas, K. J. (2004). Computer models of early land plant evolution. *Annual Review of Earth and Planetary Science* **32**: 47–66.

Norell, M. A. and X. Xing (2005). Feathered dinosaurs. *Annual Review of Earth and Planetary Sciences* **33**: 277–99.

Nummela, S., et al. (2004). Eocene evolution of whale hearing. *Nature* **430**: 776–8.

O'Connor, P. M. and L. P. A. M. Claessens (2005). Basic avian pulmonary design and flow-through ventilation in non-avian theropod dinosaurs. *Nature* **436**: 253–6.

Olson, E. C. and R. L. Miller (1958). *Morphological Integration*. Chicago, IL: Chicago University Press.

Orr, H. A. (2000). Adaptation and the cost of complexity. *Evolution* **54**: 113–20.

Ostrom, J. H. (1973). The ancestry of birds. *Nature* **242**: 136.

Ostrom, J. H. (1976). *Archaeopteryx* and the ancestry of birds. *Biological Journal of the Linnean Society* **8**: 91–182.

Ostrom, J. H. (1979). Bird flight: how did it begin? *American Scientist* **67**(1): 46–56.

Owen, R. (1876). *Descriptive and illustrated catalogue of the Fossil Reptilia of South Africa in the collection of the British Museum*. London.

Padian, K. and A. de Ricqlès (2009). L'origin et l'évolution des oiseaux: 35 années de progrès. *Comptes Rendus Palevol* **8**: 257–80.

Padian, K. and J. R. Horner (2004). Dinosaur physiology. In D. B. Weishampel, P. Dodson, and H. Osmólska (eds), *The Dinosauria*. Berkeley, CA: University of California Press, 660–71.

Page, R. D. M. and E. C. Holmes (eds), (1998). *Molecular Evolution: A Phyletic Approach*. Oxford: Blackwell Science.

Palmer, C. (2014). The aerodynamics of gliding flight and its application to the arboreal flight of the Chinese feathered dinosaur *Microraptor*. *Biological Journal of the Linnean Society* **113**: 828–35.

Parham, J. F., et al. (2012). Best practice for justifying fossil calibrations. *Systematic Biology* **61**(2): 346–59.

Parkhaev, P. Y. (2007). The Cambrian 'basement' of gastropod evolution. In P. Vickers-Rich and P. Komarower (eds), *The Rise and Fall of the Ediacaran Biota*, Special Publication of the Geological Society No. 286. London: The Geological Society of London, 415–21.

Parrington, F. R. (1971). On the Upper Triassic mammals. *Philosophical Transactions of the Royal Society B* **261**: 231–72.

Patterson, C. (1981). Significance of fossils in determining evolutionary relationships. *Annual Review of Ecology and Systematics* **12**: 195–23.

Patterson, C. and D. E. Rosen (1977). Review of the ichthyodectiform and other Mesozoic teleost fishes and the theory and practice of classifying fossils. *Bulletin of the American Museum of Natural History* **158**: 81–172.

Peel, J. S. (1991). Functional morphology of the class Helicionelloida nov., and the early evolution of the Mollusca. In M. Simonetta and S. Conway Morris (eds), *The Early Evolution of Metazoa and the Significance of Problematic Taxa*. Cambridge: Cambridge University Press, 157–77.

Peel, J. S. (2006). Scaphopodization in Palaeozoic molluscs. *Palaeontology* **49**(6): 1357–64.

Peel, J. S. (2010). Articulated hyoliths and other fossils from the Sirius Passet Lagerstätte (early Cambrian) of North Greenland. *Bulletin of Geosciences* **85**(3): 385–94.

Pennisi, E. (2013). The man who bottled evolution. *Science* **342**: 790–3.

Peter, I. S. and E. H. Davidson (2011). Evolution of gene regulatory networks controlling body plan development. *Cell* **144**: 970–84.

Peters, S. E. and R. R. Gaines (2012). Formation of the 'Great Unconformity' as a trigger for the Cambrian explosion. *Nature* **484**: 363–6.

Peterson, K. J., et al. (2008). The Ediacaran emergence of bilaterians: congruence between the genetic and geological fossil records. *Philosophical Transactions of the Royal Society B* **363**: 1435–43.

Petronis, A. (2010). Epigenetics as a unifying principle in the aetiology of complex traits and diseases. *Nature* **465**: 721–7.

Petryshyn, V. A., et al. (2013). Petrographic analysis of new specimens of the putative microfossil *Vernanimalcula guizhouens* (Doushantuo Formation, South China). *Precambrian Research* **225**: 58–66.

Pick, L. and A. Heffer (2012). *Hox* gene evolution: multiple mechanisms contributing to evolutionary novelties. *Annals of the New York Academy of Sciences* **1256**: 15–32.

Pie, M. and J. S. Weitz (2005). A null model of morphospace occupation. *The American Naturalist* **166**(1): E1–E13.

Pierce, S. E., et al. (2012). Three-dimensional limb joint mobility in the early tetrapod *Ichthyostega*. *Nature* **486**: 523–6.

Pierce, S. E., et al. (2013). Historical perspectives on the evolution of tetrapodomorph movement. *Integrative and Comparative Biology* **53**(2): 209–23.

Pigliucci, M. (2010). Phenotypic plasticity. In M. Pigliucci and G. B. Müller (eds), *Evolution—The Extended Synthesis*. Cambridge, MA: MIT Press, 355–78.

Pigliucci, M. and G. B. Müller (2010). Elements of an extended synthesis. In M. Pigliucci and G. B. Müller (eds), *Evolution—The Extended Synthesis*. Cambridge, MA: MIT Press, 3–18.

Pisani, D., et al. (2012). Resolving phylogenetic signal from noise when divergence is rapid: a new approach to the old problem of echinoderm class relationships. *Molecular Phylogenetics and Evolution* **62**: 27–34.

Platnick, N. (1979). Philosophy and the transformation of cladistics. *Systematic Zoology* **28**: 537–46.

Pontzer, H., et al. (2009). Biomechanics of running indicates endothermy in bipedal dinosaurs. *PLoS One* **4**(11): e7783.

Porter, S. M. (2008). Skeletal microstructure indicates chancelloriids and halkieriids are closely related. *Palaeontology* **51**(4): 865–79.

Porto, A., et al. (2009). The evolution of modularity in the mammalian skull I: morphological integration patterns. *Evolutionary Biology* **36**: 118–35.

Puttick, M. N., et al. (2014). High rates of evolution preceded the origin of birds. *Evolution* **68**: 1497–510.

Ramírez, M. J. (2007). Homology as a parsimony problem: a dynamic approach for morphological data. *Cladistics* **23**: 588–612.

Rauhut, O. W. M. (2003). The interrelationships and evolution of basal theropod dinosaurs. *Special Papers in Palaeontology* **69**: 1–213.

Rees, P. M., et al. (2002). Permian phytographic patterns and climate data/model comparisons. *Journal of Geology* **110**: 1–31.

Reichert, K. B. (1837). Ueber die Visceralbogen der Wirbelthiere im Allgemeinen und deren Metamorphosen bei den Vögeln und Säugethieren. *Archiv für Anatomie, Physiologie und wissenschaftliche Medicin* (1837): 120–2.

Reidl, R. (1977). A systems analysis approach to macroevolutionary phenomena. *Quarterly Review of Biology* **52**: 351–70.

Reidl, R. (1978). *Order in Living Organisms*. Chichester: John Wiley & Sons.

Reisz, R. R., et al. (2009). *Eothyris* and *Oedaleops*: do these early Permian synapsids from Texas and New Mexico form a clade? *Journal of Vertebrate Paleontology* **29**(1): 39–47.

Retallack, G. J. (1994). Were the Ediacaran fossils lichens? *Paleobiology* **20**(4): 528–44.

Retallack, G. J. (2013). Ediacaran life on land. *Nature* **493**: 89–92.

Richardson, M. K., et al. (2009). Heterochrony in limb evolution: developmental mechanisms and natural selection. *Journal of Experimental Zoology (Molecular and Developmental Evolution)* **312B**: 639–64.

Riddihough, G. and L. M. Zahn (2010). What is epigenetics? *Science* **330**: 611.

Rieppel, O. (1996). Miniaturization in tetrapods: consequences for skull morphology. In P. J. Miller (ed.), *Symposium on Miniature Vertebrates—The Implications of Small Body Size*. London: Zoological Society of London, 47–61.

Rieppel, O. (2001). Preformationist and epigenetic biases in the history of the morphological character concept. In G. P. Wagner (ed.), *The Character Concept in Evolutionary Biology*. San Diego, CA: Academic Press, 59–77.

Rieppel, O. (2001a). Turtles as hopeful monsters. *BioEssays* **23**: 987–91.

Rieppel, O. and M. Kearney (2002). Similarity. *Biological Journal of the Linnean Society* **75**: 59–82.

Rieppel, O. and R. R. Reisz (1999). The origin and early evolution of turtles. *Annual Review of Ecology and Systematics* **30**: 1–22.

Romer, A. S. (1958). Tetrapod limbs and early tetrapod life. *Evolution* **12**: 365–9.

Roseman, C. C., et al. (2009). Phenotypic integration without modularity: testing hypotheses about the distribution of pleiotropic quantitative trait loci in a continuous space. *Evolutionary Biology* **36**: 282–91.

Rota-Stabelli, O., et al. (2010). Ecdysozoan mitogenomics: evidence for a common origin of the legged invertebrates: the Panarthropoda. *Genome Biology and Evolution* **2**: 425–40.

Royer, D. L., et al. (2004). CO_2 as a primary driver of Phanerozoic climate. *GSA Today* **14**: 4–10.

Rozhnov, S. V. (2009). The role of heterochrony in the establishment of body plan in higher echinoderm taxa. *Biology Bulletin* **36**(2): 117–27.

Ruben, J. A., et al. (1997). Lung structure and ventilation in theropod dinosaurs and early birds. *Science* **278**: 1267–70.

Ruben, J. A., et al. (1999). Pulmonary function and metabolic physiology of theropod dinosaurs. *Science* **283**: 514–16.

Ruben, J. A., et al. (2012). The evolution of mammalian endothermy. In A. Chinsamy-Turan (ed.), *Forerunners of*

Mammals: Radiation, Histology, Biology. Bloomington, IN: Indiana University Press.

Rundell, R. J. and B. S. Leander (2010). Masters of miniaturization: convergent evolution among interstitial eukaryotes. *BioEssays* **32**: 430–7.

Runnegar, B. (2011). Once again: is *Nectocaris pteryx* a stem-group cephalopod? *Lethaia* **44**: 373.

Runnegar, B. and J. Pojeta (1972). The earliest bivalves and their Ordovician descendants. *American Malacological Bulletin* **9**(2): 117–22.

Runnegar, B., et al. (1979). New species of the Cambrian and Ordovician chitons *Matthevia* and *Chelodes* from Wisconsin and Queensland: evidence for the early history of polyplacophoran molluscs. *Journal of Paleontology* **53**(6): 1374–94.

Ruta, M., et al. (2006). Evolutionary patterns in early tetrapods. I. Rapid initial diversification followed by decrease in rates of character change. *Proceedings of the Royal Society B* **273**: 2113–18.

Ryan, J. F., et al. (2013). The genome of the ctenophore *Mnemiopsis leidyi* and its implications for cell type evolution. *Science* **342**: 1242–592.

Salazar-Ciudad, I. (2009). Looking at the origin of phenotypic variation from pattern formation genetic networks. *Journal of Biosciences* **34**(4): 573–87.

Sanger, T. J., et al. (2011). Roles for modularity and constraint in the evolution of cranial diversity among *Anolis* lizards. *Evolution* **66**(5): 1525–42.

Säve-Söderbergh, G. (1932). Preliminary note on Devonian stegocephalians from East Greenland. *Meddelelser om Grønland* **98**(3): 1–211.

Schachner, E. R., et al. (2009). Evolution of the respiratory system in nonavian theropods: evidence from rib and vertebral morphology. *The Anatomical Record* **292**: 1501–13.

Schlosser, G. (2002). Modularity and the units of evolution. *Theory in Biosciences* **121**: 1–80.

Schlosser, G. (2004). The role of modules in development and evolution. In G. Schlosser and G. P. Wagner (eds), *Modularity in Development and Evolution*. Chicago, IL: Chicago University Press, 519–82.

Schlosser, G. and G. P. Wagner (eds), (2004). *Modularity in Development and Evolution*. Chicago, IL: Chicago University Press.

Scholtz, G. (2010). Deconstructing morphology. *Acta Zoologica* **91**: 44–63.

Schrag, D. P., et al. (2013). Authigenic carbonate and the history of the global carbon cycle. *Science* **339**: 540–3.

Schwenk, K. and G. P. Wagner (2001). Function and the evolution of phenotypic stability: connecting pattern to process. *American Zoologist* **41**: 522–63.

Segal, E., et al. (2003). Module networks: identifying regulatory modules and their condition-specific regulators from gene expression data. *Nature Genetics* **14**(2): 166–76.

Seilacher, A. (2007). The nature of vendobionts. In P. Vickers-Rich and P. Komarower (eds), *The Rise and Fall of the Ediacaran Biota*. Special Publication No. 286. London: Geological Society of London, 387–97.

Sellers, A. I., et al. (2009). Virtual palaeontology: gait reconstruction of extinct vertebrates using high performance computing. *Palaeontologica Electronica* **12**(3): No.13A.

Senter, P. (2007). A new look at the phylogeny of Coelurosauria (Dinosauria: Theropoda). *Journal of Systematic Palaeontology* **5**(4): 429–63.

Sereno, R. C. and A. B. Arcucci (1994). Dinosaurian precursors from the Middle Triassic of Argentina. *Journal of Vertebrate Paleontology* **14**: 43–73.

Sereno, P. C., et al. (2009). Tyrannosaurid skeletal design first evolved at small body size. *Science* **326**: 418–22.

Shedlock, A. M. and S. V. Edwards (2009). Amniotes (Amniota). In S. B. Hedges and S. Kumar (eds), *The Timetree of Life*. Oxford: Oxford University Press, 375–9.

Shimeld, S. M. and P. W. H. Holland (2000). Vertebrate innovations. *Proceedings of the National Academy of Sciences* **97**: 4449–52.

Shu, D., et al. (1996). Reinterpretation of *Yunnanozoon* as the earliest hemichordate. *Nature* **380**: 428–30.

Shu, D.-G., et al. (2001). Primitive deuterostomes from the Chengjiang Lagerstätte (Lower Cambrian, China). *Nature* **414**: 419–24.

Shu, D. -G., et al. (2003). A new species of yunnanozoan with implications for deuterostome evolution. *Science* **299**: 1380–84.

Shu, D.-G., et al. (2004). Ancestral echinoderms from the Chengjiang deposits of China. *Nature* **430**: 422–28.

Shu, D.-G., et al. (2010). The earliest history of the deuterostomes: the importance of the Chengjiang Fossil-Lagerstätte. *Proceedings of the Royal Society B* **277**: 165–74.

Shubin, N. H. and E. B. Daeschler (2014). Pelvic girdle and fin of *Tiktaalik roseae*. *Proceedings of the National Academy of Sciences* **111**: 893–9.

Shubin, N. H. and M. D. David (2004). Modularity in the evolution of vertebrate appendages. In G. Schlosser and G. P. Wagner (eds), *Modularity in Evolution and Development*. Chicago, IL: Chicago University Press, 429–40.

Shubin, N. H., et al. (2006). The pectoral fin of *Tiktaalik rosaea* and the origin of the tetrapod limb. *Nature* **440**: 764–71.

Shubin, N., et al. (2009). Deep homology and the origins of evolutionary novelty. *Nature* **457**: 818–23.

Shun-Ichiro, N. (2011). On the integrated framework of species concepts: Mayden's hierarchy of species concepts and de Queiroz's on a unified concept of species. *Journal of Zoological Systematics and Evolutionary Research* **49**(3): 177–84.

Sigwart, J. D. and M. D. Sutton (2007). Deep molluscan phylogeny: synthesis of palaeontological and neontological data. *Proceedings of the Royal Society B* **274**: 2413–19.

Simpson, G. G. (1944). *Tempo and Mode in Evolution.* New York, NY: Columbia University Press.

Simpson, G. G. (1945). Principles of classification and a classification of mammals. *Bulletin of the American Museum of Natural History* **85**: 1–350.

Simpson, G. G. (1961). *Principles of Animal Taxonomy.* New York, NY: Columbia University Press.

Smith, A. B. (1994). *Systematics and the Fossil Record: Documenting Evolutionary Patterns.* Oxford: Blackwell Scientific.

Smith, A. B. (2004). Echinoderm roots. *Nature* **430**: 411–12.

Smith, A. B. (2005). The pre-radial history of echinoderms. *Geological Journal* **40**: 255–80.

Smith, A. B. and M. Reich (2013). Tracing the evolution of the holothurian body plan through stem-group fossils. *Biological Journal of the Linnean Society* **109**: 670–81.

Smith, M. R. (2012). Mouthparts of the Burgess Shale fossils *Odontogriphus* and *Wiwaxia*: implications for the ancestral molluscan radula. *Philosophical Transactions of the Royal Society B* **279**: 4287–95.

Smith, M. R. (2013). Nectocaridid ecology, diversity, and affinity: early origin of a cephalopod-like body plan. *Paleobiology* **39**(2): 297–321.

Smith, M. R. (2013a). Ontogeny, morphology and taxonomy of the soft-bodied Cambrian 'mollusc' *Wiwaxia*. *Palaeontology* **57**: 215–29.

Smith, M. R. and J.-B. Caron (2010). Primitive soft-bodied cephalopods from the Cambrian. *Nature* **465**: 469–72.

Smith, M. R. and J. Ortega-Hernandez (2014). *Hallucigenia*'s onychophoran-like claws and the case for Tactopoda. *Nature* **514**: 363–6.

Sober, E. (1988). *Reconstructing the Past: Parsimony, Evolution, and Inference.* Cambridge, MA: MIT Press.

Sørensen, M. V., et al. (2008). New data from an enigmatic phylum: evidence from molecular sequence data supports a sister-group relationship between Loricifera and Nematomorpha. *Journal of Zoological, Systematic and Evolutionary Research* **46**(3): 231–39.

Spalding, M., et al. (2009). Relationships of Cetacea (Artiodactyla) among mammals: increased taxon sampling alters interpretations of key fossils and character evolution. *PLoS One* **4**(9, e7062): 1–14.

Stein, M. (2010). A new arthropod from the Early Cambrian of North Greenland, with a 'great appendage'-like antennula. *Zoological Journal of the Linnean Society* **158**: 477–500.

Sun, Y., et al. (2012). Lethally hot temperatures during the Early Triassic greenhouse. *Science* **338**: 366–70.

Sutton, M. D., et al. (2004). Computer reconstruction and analysis of the vermiform mollusc *Acaenoplax hayae* from the Herefordshire Lagerstätte (Silurian, England), and implications for molluscan phylogeny. *Palaeontology* **47**(2): 293–318.

Sutton, M. D., et al. (2012). A Silurian armoured aplacophoran and implications for molluscan phylogeny. *Nature* **490**: 94–7.

Svensson, E. I. and R. Calsbeek (eds), (2012). *The Adaptive Landscape in Evolutionary Biology.* Oxford: Oxford University Press.

Swalla, B. J. and A. B. Smith (2008). Deciphering deuterostome phylogeny: molecular, morphological and palaeontological perspectives. *Philosophical Transactions of the Royal Society B* **363**: 1557–68.

Takezaki, N., et al. (2004). The phylogenetic relationship of tetrapod, coelacanth, and lungfish revealed by the sequences of forty-four nuclear genes. *Molecular Biology and Evolution* **21**(8): 1512–24.

Tanaka, M. and C. Tickle (2007). The development of fins and limbs. In B. K. Hall (ed.), *Fins into Limbs.* Chicago, IL: Chicago University Press, 65–78.

Telford, M. J., et al. (2008). The evolution of the Ecdysozoa. *Philosophical Transactions of the Royal Society B* **363**: 1529–37.

Thewissen, J. G. M., et al. (1994). Fossil evidence for the origin of aquatic locomotion in archaeocete whales. *Science* **263**: 210–12.

Thewissen, J. G. M., et al. (1996). *Ambulocetus natans,* an Eocene cetacean (Mammalia) from Pakistan. *Courier Forschungsinstitut Senckenberg* **191**: 1–86.

Thewissen, J. G. M., et al. (2001). Skeletons of terrestrial cetaceans and the relationship of whales to artiodactyls. *Nature* **413**: 277–81.

Thewissen, J. G. M., et al. (2007). Shales originated from aquatic artiodactyls in the Eocene epoch of India. *Nature* **450**: 1190–5.

Thewissen, J. G. M., et al. (2011). Evolution of dental wear and diet during the origin of whales. *Paleobiology* **374**(4): 655–69.

Thomas, R. D. K. and W.-E. Reif (1993). The skeleton space: a finite set of organic designs. *Evolution* **47**: 341–60.

Thomason, J. J. and A. P. Russell (1986). Mechanical factors in the evolution of the mammalian secondary palate: a theoretical analysis. *Journal of Morphology* **189**: 199–213.

Turner, A. H., et al. (2007). A basal dromaeosaurid and size evolution preceding avian flight. *Science* **317**: 1378–81.

Turner, D. (2005). Local underdetermination in the historical sciences. *Philosophy of Science* **72**: 209–30.

Turner, D. D. (2009). How much can we know about the causes of evolutionary trends? *Biology and Philosophy* **24**: 341–57.

Turner, S., et al. (2010). False teeth: conodont-vertebrate phylogenetic relationships revisited. *Geodiversitas* **32**(4): 545–94.

Twitchett, J. W. (2006). The palaeoclimatology, palaeoecology, and palaeoenvironmental analysis of mass extinction events. *Palaeogeography, Palaeoclimatology, Palaeoecology* **232**: 190–213.

Tziperman, E., et al. (2011). Biologically induced initiation of the Neoproterozoic snowball-Earth events. *Proceedings of the National Academy of Sciences* **108**(37): 15091–96.

Uhen, M. D. (2007). The terrestrial to aquatic transition in Cetacea. In J. S. Anderson and H.-D. Sues (eds), *Major Transitions in Vertebrate Evolution*. Bloomington, IN: Indiana University Press, 392–408.

Valentine, J. W. (2004). *On the Origin of Phyla*. Chicago, IL: Chicago University Press.

Varricchio, D. J. (2008). Avian paternal care had dinosaur origin. *Science* **322**: 1826–8.

Veizer, J. (1988). The Earth and its life: systems perspective. *The Origin of Life and the Evolution of the Biosphere* **18**: 13–39.

Vendrasco, M. J. and B. Runnegar (2004). Late Cambrian and Early Ordovician stem group chitons (Mollusca: Polyplacophora) from Utah and Missouri. *Journal of Paleontology* **78**(4): 675–89.

Vickers-Rich, P. and P. Komarower (eds), (2007). *The Rise and Fall of the Ediacaran Biota*. Special Publications. London: The Geological Society.

Vinther, J. (2009). The canal system in sclerites of Lower Cambrian *Sinosachites* (Halkieriidae: Sachitida): significance for the molluscan affinities of the sachitids. *Palaeontology* **52**(4): 689–712.

Vinther, J. (2015). The origin of molluscs. *Palaeontology* **58**: 19–34.

Vinther, J. and C. Nielsen (2005). The Early Cambrian *Halkeria* is a mollusc. *Zoologica Scripta* **34**: 81–9.

Vinther, J., et al. (2012). A molecular palaeontological hypothesis for the origin of aplacophoran molluscs and their derivation from chiton-like ancestors. *Proceedings of the Royal Society B* **279**: 1259–68.

Vinther, J., et al. (2014). A filter feeding anomalocaridid from the Early Cambrian. *Nature* **507**: 496–9.

Vogt, L., et al. (2010). The linguistic problem of morphology: structure versus homology and the standardization of morphological data. *Cladistics* **26**: 301–25.

Vorobyeva, E. I. and H.-P. Schultze (1991). Description and systematics of panderichthyid fishes with comments on their relationships to tetrapods. In H.-P. Schultze and L. Trueb (eds), *Origins of the Higher Groups of Tetrapods: Controversy and Consensus*. New York, NY: Cornell University Press, 68–109.

Waddington, C. H. (1953). Genetic assimilation of an acquired character. *Evolution* **7**: 118–26.

Wagner, G. P. (1988). The influence of variation and of developmental constraints on the rate of multivariate phenotypic evolution. *Journal of Evolutionary Biology* **1**: 45–66.

Wagner, G. P., ed. (2001). *The Character Concept in Evolutionary Biology*. San Diego, CA: Academic Press.

Wagner, G. P. and L. Altenberg (1996). Complex adaptations and the evolution of evolvability. *Evolution* **50**(3): 967–76.

Wagner, G. P. and J. Zhang (2011). The pleiotropic structure of the genotype–phenotype map: the evolvability of complex organisms. *Nature Reviews Genetics* **12**: 204–13.

Wagner, G. P., et al. (2007). The road to modularity. *Nature Reviews Genetics* **8**: 921–31.

Wainwright, P. C. (2007). Functional versus morphological diversity in macroevolution. *Annual Review of Ecology, Evolution and Systematics* **38**: 381–401.

Wainwright, P. C., et al. (2012). The evolution of pharyngeognathy: a phylogenetic and functional appraisal of the pharyngeal jaw key innovation in labroid fishes and beyond. *Systematic Biology* **61**(6): 1001–27.

Walcott, C. D. (1911). Cambrian geology and paleontology. II. Middle Cambrian annelids. *Smithsonian Miscellaneous Collections* **67**: 109–44.

Walker, J. A. (2007). A general model of functional constraints on phenotypic evolution. *The American Naturalist* **170**(5): 681–89.

Waloszek, D., et al. (2005). Early Cambrian arthropods—new insights into arthropod head and structural evolution. *Arthropod Structure and Development* **34**: 189–205.

Waloszek, D., et al. (2007). Evolution of cephalic feeding structures and the phylogeny of arthropods. *Palaeogeography, Palaeoclimatology, Palaeoecology* **254**: 273–87.

Wang, Z., et al. (2010). Genomic patterns of pleiotropy and the evolution of complexity. *Proceedings of the National Academy of Sciences* **107**(42): 18034–9.

Watson, D. M. S. (1914). *Eunotosaurus africanus* Seeley, and the ancestry of Chelonia. *Proceedings of the Zoological Society of London for 1914*: 1011–20.

Weisbecker, V., et al. (2008). Ossification heterochrony in the therian postcranial skeleton and the marsupial-placental dichotomy. *Evolution* **62**(8): 2027–41.

Weishampel, D. B., et al. (eds), (2004). *The Dinosauria*, 2nd edition. Berkeley, CA: University of California Press.

West-Eberhard, M. J. (2003). *Developmental Plasticity and Evolution*. Oxford, Oxford University Press.

White, L. L. (1965). *Internal Factors in Evolution*. London: Tavistock Publications Ltd.

Whiteaves, J. F. (1883). Recent discoveries of fossil fishes in the Devonian rocks of Canada. *The American Naturalist* **17**: 158–64.

Williams, C. G. (1992). *Natural Selection: Domains, Levels, and Challenges*. Oxford: Oxford University Press.

Willmer, P. (1990). *Invertebrate Relationships: Patterns in Animal Evolution*. Cambridge: Cambridge University Press.

Wills, M. A. (1998). Crustacean disparity through the Phanerozoic: comparing morphological and stratigraphic data. *Biological Journal of the Linnean Society* **65**: 455–500.

Wills, M. A., et al. (1994). Disparity as an evolutionary index: a comparison of Cambrian and recent arthropods. *Paleobiology* **20**: 93–130.

Wilson, N. G., et al. (2010). Assessing the molluscan hypothesis Serialia (Monoplacophora + Polyplacophora) using novel molecular data. *Molecular Phylogenetics and Evolution* **54**: 187–93.

Witmer, L. M. (1991). Perspectives on avian origins. In H.-P. Schultze and L. Trueb (eds), *Origins of the Higher Groups of Tetrapods: Controversy and Consensus*. Ithaca, NY: Comstock Publishing Associates, 427–66.

Witzmann, F. (2010). Morphological and histological changes of dermal scales during the fish-to-tetrapod transition. *Acta Zoologica* **92**: 281–302.

Wolpert, L., et al. (2011). *Principles of Development*. Oxford: Oxford University Press.

Wright, S. (1932). The roles of mutation, inbreeding, cross-breeding and selection in evolution. *Proceedings of the Sixth Annual Congress of Genetics* **1**: 356–66.

Xian-Guang, H., et al. (2002). New evidence on the anatomy and phylogeny of the earliest vertebrates. *Philosophical Transactions of the Royal Society B* **269**: 1865–69.

Xian-Guang, H., et al. (2006). The Lower Cambrian *Phlogites* Luo & Hu reconsidered. *GFF (Journal of the Geological Society of Sweden)* **128**: 47–51.

Xiao, S. (2013). Muddying the waters. *Nature* **493**: 28–9.

Xiao, S. and M. Laflamme (2009). On the eve of animal radiation: phylogeny, ecology and evolution of the Ediacara biota. *Trends in Ecology and Evolution* **24**: 31–40.

Xing, X., et al. (2004). Basal tyrannosaurids from China and evidence for protofeathers in tyrannosauroids. *Nature* **431**: 680–4.

Xu, X. and M. A. Norell (2004). A new troodontid dinosaur from China with avian-like sleeping posture. *Nature* **431**: 838–41.

Xu, X. and M. A. Norell (2006). Non-avian fossils from the Lower Cretaceous Jehol Group of western Liaoning, China. *Geological Journal* **41**: 419–37.

Xu, X., et al. (2003). Four-winged dinosaurs from China. *Nature* **421**: 335–40.

Xu, X., et al. (2009). A new feathered maniraptoran dinosaur fossil that fills a morphological gap in avian origin. *Chinese Science Bulletin* **54**(3): 430–5.

Xu, X., et al. (2010). Pre-*Archaeopteryx* coelurosaurian dinosaurs and their implications for understanding avian origins. *Chinese Science Bulletin* **55**: 3971–7.

Xu, X., et al. (2011). An *Archaeopteryx*-like theropod from China and the origin of Avialae. *Nature* **475**: 465–70.

Yano, T. and K. Tamura (2013). The making of differences between fins and limbs. *Journal of Anatomy* **222**: 100–13.

Young, N. M. and B. Hallgrímsson (2005). Serial homology and the evolution of mammalian limb covariation structure. *Evolution* **59**(12): 2691–704.

Zardoya, R., et al. (1998). Searching for the closest living relative(s) of tetrapods through evolutionary analysis of mitochondrial and nuclear data. *Molecular Biology and Evolution* **15**(5): 506–17.

Zelditch, M. L. and W. L. Fink (1996). Heterochrony and heterotopy: stability and innovation in the evolution of form. *Paleobiology* **22**(2): 241–54.

Zelditch, M. L., et al. (2008). Modularity of the rodent mandible: integrating bones, muscles, and teeth. *Evolution and Development* **10**(6): 756–68.

Zhang, F. C. and J. O. Farlow (2001). Flight capability and habits of *Confuciusornis*. In J. Gauthier and L. F. Gall (eds), *New Perspectives on the Origin and Early Evolution of Birds*. New Haven, CT: Yale University Press, 237–54.

Zhang, F. C., et al. (2002). A juvenile coelurosaurian theropod from China indicates arboreal habits. *Naturwissenschaften* **89**(9): 394–98.

Zhang, F. C., et al. (2008). A bizarre Jurassic maniraptoran from China with elongate ribbon-like feathers. *Nature* **455**(7216): 1105–08.

Zhang, X.-G. and D. E. G. Briggs (2007). The nature and significance of the appendages of *Opabinia* from the Middle Cambrian Burgess Shale. *Lethaia* **40**: 161–73.

Zhang, X.-L., et al. (2008). Cambrian Burgess Shale-type Lagerstätte in South China: distribution and significance. *Gondwana Research* **14**: 255–62.

Zhou, Z. and F. Zhang (2003). *Jeholornis* compared to *Archaeopteryx*, with a new understanding of the earliest avian evolution. *Naturwissenschaften* **90**: 220–5.

Zhou, Z. G. and F. C. Zhang (2002). A long-tailed, seed-eating bird from the Early Cretaceous of China. *Nature* **418**: 405–9.

Index

Note: References to figures are indicated by '*f*'.